# MacRoberts on Scottish Building Contracts

# MacRoberts on Scottish Building Contracts

**MacRoberts**
Solicitors

**Foreword by**
**The Rt. Hon. The Lord Hope of Craighead**

Blackwell
Science

Editorial Offices:
Osney Mead, Oxford OX2 0EL
24 John Street, London WC1N 2BL
23 Ainslie Place, Edinburgh EH3 6AJ
350 Main Street, Malden
MA 02148 5018, USA
54 University Street, Carlton
Victoria 3053, Australia
10, rue Casimir Delavigne
75006 Paris, France

Other Editorial Offices:

Blackwell Wissenschafts-Verlag GmbH
Kurfürstendamm 57
10707 Berlin, Germany

Blackwell Science KK
MG Kodenmacho Building
7-10 Kodenmacho Nihombashi
Chuo-ku, Tokyo 104, Japan



First published 1999

Set in 10/12pt Palatino
by DP Photosetting, Aylesbury, Bucks
Printed and bound in Great Britain by
MPG Books Ltd, Bodmin, Cornwall



DISTRIBUTORS

Marston Book Services Ltd
PO Box 269
Abingdon
Oxon OX14 4YN
(*Orders:* Tel: 01235 465500
Fax: 01235 465555)

USA
Blackwell Science, Inc.
Commerce Place
350 Main Street
Malden, MA 02148 5018
(*Orders:* Tel: 800 759 6102
781 388 8250
Fax: 781 388 8255)

Canada
Login Brothers Book Company
324 Saulteaux Crescent
Winnipeg, Manitoba R3J 3T2
(*Orders:* Tel: 204 837-2987
Fax: 204 837-3116)

Australia
Blackwell Science Pty Ltd
54 University Street
Carlton, Victoria 3053
(*Orders:* Tel: 03 9347 0300
Fax: 03 9347 5001)

A catalogue record for this title is available from the British Library

ISBN 0-632-03411-4

Library of Congress
Cataloging-in-Publication Data
MacRoberts on Scottish building contracts/
foreword by the Rt. Hon. the Lord Hope of Craighead.
p. cm.
Includes index.
ISBN 0-632-03411-4 (pbk.)
1. Construction contracts - Scotland.
KDC501.M33 1999
343.411'078624 - dc21 99-34837
CIP

For further information on
Blackwell Science, visit our website:
www.blackwell-science.com

*To Jim Arnott, without whom the practice of construction law in Scotland, and the firm of MacRoberts, would not be where they are today.*

# Contents

# Foreword

The Rt. Hon. The Lord Hope of Craighead

The complexity of modern building contract law is the product of a familiar phenomenon - the desire of those who regularly enter into contractual relationships to set out in a standard form the basis of the engagement by which goods and services are to be supplied or obtained and paid for. No standard form can expect to be in use for very long before giving rise to disputes. Situations will arise which have not been provided for, words will be shown to have failed to express the meaning which one or other of the parties expected of them and experience will have demonstrated the need for greater precision on the one hand or for more flexibility on the other. Litigation may have to be resorted to, and the case law which this generates will suggest that changes need to be made. So revised forms will be issued and new standard forms will be devised. The parties themselves may wish to modify the standard forms to suit their own particular circumstances.

Standard forms of this kind for use in building contracts have been in circulation since at least the third quarter of the nineteenth century. The Royal Institute of British Architects did much of the early work in the field. Early versions of the Standard Form of Building Contract were issued under its name. Now it is only one of the many bodies that are represented on the Joint Contracts Tribunal (the JCT), which was set up in England in the early 1960s to take over responsibility for revising and re-issuing later versions of the Standard Form. It had already been appreciated that there was a need for a standard form appropriate for use in Scotland. In 1915 a form known as the Scottish National Building Code was issued. It was replaced by the Regulations and General Conditions of Contract for Building Works in Scotland dated 1 September 1954. But it was not until the Scottish Building Contracts Committee (the SBCC) was set up by various Scottish bodies interested in the carrying out of building and engineering work in Scotland, as the Scottish equivalent of the JCT, that it became possible to issue Scottish versions of the various forms in use in England, to keep them under regular review and to issue revised versions when this was required.

The progress which has been made by the SBCC in developing a series of standard forms for use in Scotland has in turn demonstrated the need for a Scottish textbook. As J.M. Arnott and W.J. Wolffe observed in *The Laws of Scotland, Stair Memorial Encyclopaedia, vol. 3, Building Contracts* (1994), the practitioner in this field cannot avoid using the standard English textbooks. But he must use them with care because they are written from the English standpoint. Matters have now reached the state where satisfactory guidance as to the forms currently in use in Scotland can only be provided by a textbook which is devoted entirely to this subject, and has been written by specialists in Scots law.

MacRoberts are to be congratulated on the initiative which they have taken in the preparation of this valuable new textbook. The firm has wide experience in the law and practice of building contracts, as it has specialised in the field for many years. Their senior partner, J.M. Arnott, is Secretary and Legal Adviser to the SBCC. As one would expect from this background, the book is designed to provide practical advice. It will serve as a companion volume for use with the SBCC forms. But it also deals with general principles which are relevant to building contracts of all kinds. I wish it a long life and every success in this important task. I hope that it may be possible for it, like the forms, to be kept up to date by the issuing of revised editions at appropriate intervals.

David Hope

# Preface

The object of this book is to bring some light to the darkness of the law of Scotland in relation to building contracts. It is not meant to be a Scottish *Keating*; however it does offer a Scottish perspective on this complex, and at times esoteric, area of the law. For those involved in the construction industry in Scotland, and for their advisers, it is hoped that this book will go some way to answering some of the everyday legal issues they are confronted with.

At least in the last forty or so years, construction and engineering law has developed in Scotland very closely with the law of England - largely due to the use in both countries of common standard forms of contract and the fact that most decisions on the interpretation of such forms are decisions of English courts. However readers should be mindful of the fact that there are significant differences between the respective common laws of each jurisdiction, and in certain cases it would be inappropriate for a Scottish court to follow an earlier English decision. This must always be borne in mind when considering the weight to be placed on a decision of an English court.

There is often a tendency amongst non-lawyers to regard decisions of a court of first instance to be of binding authority. This is incorrect. A subsequent court is bound by the precedent of an earlier decision only if that earlier decision is 'in point', i.e. it is not possible to distinguish between the facts of the earlier and later cases, and is 'binding'. Whether a decision is binding depends upon the hierarchy of the courts. For example, the decisions of the House of Lords in Scottish appeals are binding on all civil courts in Scotland. However, the decision of the House of Lords in an English appeal is not binding in the Scottish courts unless the point at issue, whether in relation to legislation or to a particular clause in a contract, is exactly the same in both countries.

Where an earlier decision in point is not binding it can be treated as persuasive by a subsequent court; examples would be the decision of the House of Lords or Court of Appeal in an English case where the facts and the law coincide, or the decision of a judge sitting in the Outer House of the Court of Session in its application to a later Sheriff Court action.

Each of the individual contributors to this work deserves a mention and the thanks of the firm for their efforts. The writing was done by Lindy Patterson, Neil Kelly, Richard Barrie, Lynda Ross, David Arnott, Shona Frame, Chris Arnold, Louise Cook, Anthony Wallace and ourselves; Neil Wilson and Bryan Darroch offered helpful observations on various parts of the text; Clare McGinley and Claire Adams sorted out the table of cases; and David Flint made the preparation of the index far easier than it might have been.

We are also deeply indebted to each of the secretaries who converted the various contributions into a legible form, especially Yvonne Gallacher who also had the unenviable task of bringing the whole book together in a form that could be submitted to the publishers. Finally, we must thank Julia Burden and her colleagues at Blackwell Science for tolerating more excuses and delays than anyone could realistically be expected to tolerate and for giving us the opportunity to publish this book.

We have aimed to state the law as at 1 March 1999, however at proof stage we have endeavoured to update the text insofar as possible, subject to constraints of space. That has enabled us to have regard to the reprinting of the SBCC contracts which were published in May 1999.

David Henderson
Craig Turnbull
MacRoberts
152 Bath Street
Glasgow G2 4TB

July 1999

# Chapter 1
# Building Contracts in General

## 1.1 *Introduction*

There have been numerous books written on the subject of building contracts many of which are of a specialist nature and most of which are written from an English point of view. The purpose of this book is to provide a practical guide to building contracts governed by the law of Scotland.

The standard reference works are *Building Contracts* by D. Keating (6th ed., edited by The Hon. Sir Anthony May), *Hudson's Building and Engineering Contracts* (11th ed., edited by I.N. Duncan-Wallace), and *Emden's Construction Law* (looseleaf with update service, edited by A.J. Anderson, S. Bickford-Smith, N.E. Palmer and R. Redmond-Cooper). These books can be of considerable assistance; however, the underlying law as described in them is, in many instances, different from the law applicable in Scotland. The only recent text devoted to Scottish building contracts is a section of Volume 3 of *The Laws of Scotland – Stair Memorial Encyclopaedia* entitled *Building Contracts* by J.M. Arnott and W.J. Wolffe.

## 1.2 *Definition of a building contract*

### 1.2.1 General

For the purposes of Part II of the recently enacted Housing Grants, Construction and Regeneration Act 1996 (which we will refer to as 'the 1996 Act') the terms 'construction contract' and 'construction operations' were brought to the statute book. We will return to these terms below. However, only those construction contracts entered into after the commencement date of Part II of the 1996 Act, namely 1 May 1998, are subject to Part II and, more importantly for present purposes, only those construction contracts which relate to the carrying out of construction operations in England, Wales or Scotland are subject to Part II. Whilst the applicable law in relation to the contract is irrelevant for the purposes of Part II, in this book a Scottish building contract is a contract for the carrying out of construction operations in Scotland in respect of which the applicable law is the law of Scotland.

The definition of a building contract is not so straightforward. The construction industry is a very wide-ranging one which can consist of projects which differ enormously in nature (e.g. building or engineering), size and complexity. The phrases 'building contract' and 'construction contract' are often used interchangeably. The term 'construction contract' was introduced by s. 104 of the 1996 Act. While construction contracts are defined in considerable detail for the purposes of the 1996 Act, the definition emphasises the wide-ranging nature of building contracts.

### 1.2.2 Construction contracts under the 1996 Act

In defining what a building contract is, considerable assistance can be derived from the 1996 Act. The far-reaching effects of Part II depend upon its scope. To properly appreciate this one requires an understanding of ss 104 to 107 of the 1996 Act. Part II applies to what are termed 'construction contracts', being agreements in relation to 'construction operations'. These terms are defined respectively by ss. 104 and 105 of the 1996 Act. By virtue of s. 107 the agreement must be in writing. We will return to this below.

Section 104 provides as follows:

(1) In this Part a 'construction contract' means an agreement with a person for any of the following –

(a) the carrying out of construction operations;
(b) arranging for the carrying out of construction operations by others, whether under sub-contract to him or otherwise;
(c) providing his own labour, or the labour of others, for the carrying out of construction operations.

(2) References in this Part to a construction contract include an agreement –

(a) to do architectural, design or surveying work, or
(b) to provide advice on building, engineering, interior or exterior decoration or on the laying-out of landscape,

in relation to construction operations.

(3) References in this Part to a construction contract do not include a contract of employment (within the meaning of the Employment Rights Act 1996).

(4) The Secretary of State may by order add to, amend or repeal any of the provisions of subsection (1), (2) or (3) as to the agreements which are construction contracts for the purposes of this Part or are to be taken or not to be taken as included in references to such contracts.

No such order shall be made unless a draft of it has been laid before and approved by a resolution of each House of Parliament.

(5) Where an agreement relates to construction operations and other matters, this Part applies to it only so far as it relates to construction operations.

An agreement relates to construction operations so far as it makes provision of any kind within subsection (1) or (2).

(6) This Part applies only to construction contracts which –

(a) are entered into after the commencement of this Part, and
(b) relate to the carrying out of construction operations in England, Wales or Scotland.

(7) This Part applies whether or not the law of England and Wales or Scotland is otherwise the applicable law in relation to the contract.

Whilst s.104 (1)(a) and (c) appears self-explanatory, the term 'arranging' which appears in s.104 (1)(b) is not defined. In this regard, some assistance may be found in regulation 2(2) of the Construction (Design and Management) Regulations 1994 (hereinafter referred to as 'the CDM Regulations') where the expressions 'arrange' and 'arranges' are defined.

It will be immediately noted that the 1996 Act applies to matters beyond the carrying out of building works. It applies to architectural, design and surveying works and to advising on building, engineering, interior or exterior decoration or on the laying-out of landscape in relation to construction operations. What, precisely 'providing advice' is deemed to cover will no doubt become apparent in due course - it certainly is not clear at present. It is not, however, thought to cover the provision of legal advice in relation to disputes arising from construction operations.

The situation where an agreement deals with construction operations and other, inevitably related, matters is dealt with by s.104 (5). It does not matter whether construction operations are the primary object of the agreement, or are ancillary. Part II applies to the agreement insofar as it relates to construction operations - not to the other matters.

Part II only applies to contracts entered into after its commencement date, namely 1 May 1998. Irrespective of the applicable law of the contract, construction contracts that are performed in Great Britain are governed by Part II of the 1996 Act by virtue of s.104 (6)(b) and s.104 (7).

As will have been seen from s.104, the definition of 'construction operations' is central to Part II. This term is defined by s.105 (1), which provides as follows:

(1) In this Part 'construction operations' means, subject as follows, operations of any of the following descriptions –

(a) construction, alteration, repair, maintenance, extension, demolition or dismantling of buildings, or structures forming, or to form, part of the land (whether permanent or not);
(b) construction, alteration, repair, maintenance, extension, demolition or dismantling of any works forming, or to form, part of the land, including (without prejudice to the foregoing) walls, roadworks, power-lines, telecommunication apparatus, aircraft runways, docks and harbours, railways, inland waterways, pipe-lines, reservoirs, water-mains, wells, sewers, industrial plant and installations for purposes of land drainage, coast protection or defence;
(c) installation in any building or structure of fittings forming part of the land, including (without prejudice to the foregoing) systems of heating, lighting, air-conditioning, ventilation, power supply, drainage, sanitation, water supply or fire protection, or security or communications systems;
(d) external or internal cleaning of buildings and structures, so far as carried out in the course of their construction, alteration, repair, extension or restoration;
(e) operations which form an integral part of, or are preparatory to, or are for rendering complete, such operations as are previously described in this subsection including site clearance, earth-moving, excavation, tunnelling and boring, laying of foundations, erection, maintenance or dismantling of scaffolding, site restoration, landscaping and the provision of roadways and other access works;
(f) painting or decorating the internal or external surfaces of any building or structure.

It is perhaps unfortunate that Parliament found it necessary to provide a new term, 'construction operations' only two years after we were given the term 'construction work' by the CDM Regulations. Prior to the CDM Regulations we had the terms 'building operation' and 'work of engineering construction' from the Factories Act 1961. In the consultation document issued by The Health and Safety Commission that led to the CDM Regulations it was stated that the aim of the new term was to

> '... establish a comprehensive and logical framework which brings into scope the full range of construction work and which removes as many as possible of the anomalies and areas of doubt associated with the existing definitions ...'.

It is doubtful if this aim has been achieved.

Unfortunately, the differences between the definitions 'construction work' and 'construction operations' have left those involved in the construction industry with the unenviable task of repeatedly having to check whether one or other or both of Part II of the 1996 Act and the CDM Regulations apply to a particular project.

A line by line comparison of the two definitions is beyond the scope of this work; however, by way of example it is pertinent to note that fitting out, commissioning, renovation and upkeep are construction works by virtue of regulation 2(1)(a) of the CDM Regulations but are not construction operations under the 1996 Act. Similarly, the installation, commissioning, maintenance, repair or removal of mechanical, electrical, gas, compressed air, hydraulic, telecommunications, computer or similar services which are normally fixed within or to a structure are again construction works, by virtue of regulation 2(1)(e) of the CDM Regulations, but are not construction operations under the 1996 Act.

Section 105 (2) details a number of operations that are not construction operations for the purposes of Part II. It provides as follows:

> (2) The following operations are not construction operations within the meaning of this Part –
>
> (a) drilling for, or extraction of, oil or natural gas;
> (b) extraction (whether by underground or surface working) of minerals, tunnelling or boring, or construction of underground works, for this purpose;
> (c) assembly, installation or demolition of plant or machinery, or erection or demolition of steelwork for the purposes of supporting or providing access to plant or machinery, on a site where the primary activity is –
> (i) nuclear processing, power generation, or water or effluent treatment, or
> (ii) the production, transmission, processing or bulk storage (other than warehousing) of chemicals, pharmaceuticals, oil, gas, steel or food and drink;
> (d) manufacture or delivery to site of –
> (i) building or engineering components or equipment,
> (ii) materials, plant or machinery, or
> (iii) components for systems of heating, lighting, air-conditioning, ventilation, power supply, drainage, sanitation, water supply or fire protection, or for security or communications systems,
> except under a contract which also provides for their installation;
> (e) the making, installation and repair of artistic works, being sculptures, murals and other works which are wholly artistic in nature.

The first two exceptions, ss.105 (2) (a) and (b), relate to oil and gas and mining, both underground and opencast. The third exception, s.105 (2) (c) relates to nuclear processing, power generation, water or effluent treatment and the production, transmission, processing or bulk storage (other than warehousing) of chemicals, pharmaceuticals, oil, gas, steel or food and drink.

Notwithstanding the exceptions, it is incorrect to say that the 1996 Act does not apply to these industries. The extent of the exceptions is important. Other works relating to these industries which fall within the scope of s. 105 (1) are covered by the 1996 Act, for example, the construction of a water or effluent treatment plant is a construction operation and the 1996 Act applies. It is only the assembly, installation or demolition of plant or machinery or the erection or demolition of steelwork for the purposes of supporting or providing access to plant or machinery that is excepted. This is not a construction operation where the primary activity on the site is water or effluent treatment.

The mechanism by which s.105 can be amended is provided for by s.105(3) and (4) which provides that: –

(3) The Secretary of State may by order add to, amend or repeal any of the provisions of subsection (1) or (2) as to the operations and work to be treated as construction operations for the purposes of this Part.

(4) No order under this section shall be made unless a draft of it has been laid before and approved by a resolution of each House of Parliament.

Section 106 deals with contracts with residential occupiers and provides as follows:

(1) This Part does not apply –

(a) to a construction contract with a residential occupier, or
(c) to any other description of construction contract excluded from the operation of this Part by order of the Secretary of State.

(2) A construction contract with a residential occupier means a construction contract which principally relates to operations on a dwelling which one of the parties to the contract occupies, or intends to occupy, as his residence.

In this subsection 'dwelling' means a dwelling-house or a flat; and for this purpose

'dwelling-house' does not include a building containing a flat; and

'flat' means separate and self-contained premises constructed or adapted for use for residential purposes and forming part of a building from some other part of which the premises are divided horizontally.

(3) The Secretary of State may by order amend subsection (2).

(4) No order under this section shall be made unless a draft of it has been laid before and approved by a resolution of each House of Parliament.

In itself, s.106 is self-explanatory and merits little further comment. The powers given to the Secretary of State under this section have been used already. The Construction Contracts (Scotland) Exclusion Order 1998 was made on 6 March 1998 and came in to force with Part II of the 1996 Act on 1 May 1998. This statutory instrument is the vehicle by which private finance initiative (PFI) and other projects are partially saved from the complications of the provisions of Part II of the 1996 Act.

The order excludes from the operation of Part II certain agreements under statute, namely, construction contracts which are in respect of contributions towards constructing or improving roads under s. 48 of the Roads (Scotland) Act 1984, agreements regulating the development or use of land or relating to Crown land under ss. 75 and 246 of the Town and Country Planning (Scotland) Act 1997, agreements as to the provision of sewers etc. for new premises under s. 8 of the Sewerage (Scotland) Act 1968 or an externally financed development agreement within the meaning of s. 1 (powers of NHS Trusts to enter into agreements) of the National Health Service (Private Finance) Act 1997.

Also excluded are construction contracts entered into under the PFI within a specific defined meaning, set out in article 3 of the order. Three criteria must be satisfied for a contract to be entered into under the initiative for the purpose of the order.

First, the contract must contain a statement that it is entered into under the initiative or under a project 'applying similar principles'.

Second, the consideration under the contract must be determined, at least in part, by reference to one or more of three set criteria. These are the standards attained in the performance of a service (the provision of which is the principal purpose or one of the principal purposes for which the building or structure is constructed); the extent, rate or intensity of use of all or any part of the building or structure in question; or the right to operate any facility in connection with the building or structure in question.

Third, one of the parties to the contract *must* be either:

- a Minister of the Crown;
- a department in respect of which appropriation accounts are required to be examined and certified by or are open to the inspection of the Comptroller and Auditor General;
- a nationalised industry or other public authority listed in Schedule 4 to the National Audit Act 1983;
- a body whose accounts are subject to audit by auditors appointed by the Accounts Commission for Scotland;
- a company wholly owned by any of the foregoing bodies;

- one of the new water and sewerage authorities; or
- the board of management of a self-governing school.

Any construction contract that is entered into in connection with a PFI contract which does not have one of these bodies as a party to it *will* be regulated by Part II.

Finally, the order also excludes from the operation of Part II finance agreements and development agreements.

Section 107 of the 1996 Act relates to the application of Part II only to agreements in writing. It provides as follows:

> (1) The provisions of this Part apply only where the construction contract is in writing, and any other agreement between the parties as to any matter is effective for the purposes of this Part only if in writing.
>
> The expressions 'agreement', 'agree' and 'agreed' shall be construed accordingly.
>
> (2) There is an agreement in writing –
>
> (a) if the agreement is made in writing (whether or not it is signed by the parties),
> (b) if the agreement is made by exchange of communications in writing, or
> (c) if the agreement is evidenced in writing.
>
> (3) Where parties agree otherwise than in writing by reference to terms which are in writing, they make an agreement in writing.
>
> (4) An agreement is evidenced in writing if an agreement made otherwise than in writing is recorded by one of the parties, or by a third party, with the authority of the parties to the agreement.
>
> (5) An exchange of written submissions in adjudication proceedings, or in arbitral or legal proceedings in which the existence of an agreement otherwise than in writing is alleged by one party against another party and not denied by the other party in his response constitutes as between those parties an agreement in writing to the effect alleged.
>
> (6) References in this Part to anything being written or in writing include its being recorded by any means.

The opening words of s. 107 (1) are, to say the least, confusing when the remainder of the section is examined. Suffice to say that a construction contract does *not* have to be in writing. It can be evidenced in writing or recorded by one of the parties to the agreement (or by a third party) with authority. Having frightened many into believing that the absence of a written contract would be fatal, the 1996 Act then proceeds to provide a number of routes whereby writing is not required. Notwithstanding this, the best advice remains for there always to be a written contract in the traditional sense. That at least enables parties to focus upon their disputes, rather than disputing what their agreement is in the first place.

## 1.3 *Parties involved in a building project*

The number and identity of the persons involved in a building project can vary considerably depending on the nature and complexity of the project. At one end of the scale a private individual may engage a joiner, electrician or builder to carry out work to his home. In such an instance, the employment of anyone other than the tradesman or builder may not be necessary. At the other end of the scale major projects, such as the construction of public buildings, motorways, hotels or power stations, can involve a considerable number of parties from different professional and non-professional disciplines. It is therefore crucial to identify, particularly in a large project, the parties involved in that project, the terms of their respective appointments, the scope of each individual's involvement, and their roles within the project. The following parties are commonly involved in building projects.

### 1.3.1 Employer

The term 'employer' is used throughout this book as meaning the party for whose benefit the building works are being carried out. The term 'employer' is adopted for the reason that it is the term generally used by the Scottish Building Contract Committee (which is referred to in this book as 'the SBCC') in its standard form building contracts and associated documentation. However, other terms are in general use such as 'the owner', 'the client' or 'the authority'. It is the employer who usually assembles a team to advise him. There is no obligation or requirement upon an employer to do so. The obligations of employers under a building contract are considered below in Chapter 3. Smaller projects often do not require the involvement of anyone other than the employer and the contractor. The nature of the team varies depending upon the nature of the project. The employer's team in a large project normally consists of an architect and/or an engineer, a quantity surveyor, one or more specialist consultants and, possibly, a project manager and a clerk of works. The terms of appointment of each member of the team are very important. Each individual requires to have a clear understanding of his obligations and the inter-relationship of those obligations with those of the other parties. If the scope of each individual's obligations is not clearly defined by the employer then difficulties can arise with unnecessary overlap of work or, more importantly, in crucial issues failing to be addressed by any of the members of the team for the reason that each thought that another was responsible for the unaddressed issue. An example of the type of problem that can arise is to be found in the case of *Chesham Properties Ltd* v. *Bucknall Austin Project Management Services Ltd and Others* (1996).

This book will only address the relationship between the employer and the chosen contractor. The relationship and contractual arrangements between the employer and other individuals such as the members of his design team, for example, the architect, engineer, and specialist consultants will not be discussed in detail.

### 1.3.2 Architect

The architect is the individual who usually has overall responsibility for the project from its conception to its conclusion. In order for an architect to become a chartered architect he or she must have completed a seven year course in the design, specification and erection of buildings. When this has been completed they will appear on the register of the Architects Registration Board (ARB) and can apply for membership of the chartered professional bodies. The architect is the agent of the employer and the general law of Scotland in relation to agency applies to their actions. The scope of their actual authority depends upon the terms of their agreement with their client, the employer.

Chartered architects in Scotland can be members of the Royal Incorporation of Architects in Scotland (RIAS) and/or the Royal Institute of British Architects (RIBA). The RIAS is a charitable organisation founded in 1916 as a professional body for all chartered architects in Scotland. It produces a standard form of agreement for the appointment of an architect in Scotland. This comprises a Memorandum of Agreement (SCA/98), Scottish Conditions of Appointment, a Schedule of Services and a Schedule of Project Details. The SCA/98 sets out the principal factors to be addressed in the design process and can be adapted to suit the requirements of a specific appointment.

For the addresses of the office and bookshops see the Appendix. This also gives the website which contains details of the range of professional services offered by the RIAS. Anyone wishing to commission a chartered architect can visit the site and obtain information on over 500 practices. The website also provides information on how to select an architect. The RIAS has two specialist bookshops in Edinburgh and Glasgow.

### 1.3.3 Quantity surveyor

A quantity surveyor may be engaged by the employer or the architect to discharge specific functions. The functions tend to be of a financial nature and can include, for example, preparing bills of quantities, valuing work done and ascertaining direct loss and expense under the provisions of the building contract. Like architects, chartered surveyors are members of a professional organisation, in this case the Royal Institution of Chartered Surveyors (RICS), for details of which see the Appendix. The RICS produces a set of documents relating to the appointment of a quantity surveyor. The Scottish version is dated June 1992 and includes a Form of Agreement and Terms and Conditions. Contractors may themselves employ a quantity surveyor to price and prepare tender documents and to assist in negotiations with the employer.

### 1.3.4 Engineer

In an engineering contract the individual who stands in a similar role to that of the architect under a building contract is normally the engineer. The engineering professional organisation is the Institute of Civil Engineers (ICE). Their website contains details about the ICE including public relations, membership, specialisms and publications. The ICE does not have a separate Scottish office but has local associations, two of which are the Glasgow and West of Scotland, Highlands and Islands branch and the East of Scotland branch. Details of the local associations can be found on the ICE website.

There are also two relevant trade associations (for details see the Appendix).

The Association of Consulting Engineers (ACE) is a trade association for firms of consulting engineers. Member firms of the ACE include sole practitioners, partnerships and companies. The ACE is a member of the European Federation of Engineering Consultancy Association (EFCA) and is affiliated to the Féderation Internationale des Ingénieurs Conseils (FIDIC). The ACE publishes guidance notes, standard agreements and conditions of contract and a directory of ACE firms. A wide range of publications is available for purchase through the ACE including their own publications and publications by the BPF (British Property Federation), EFCA, FIDIC and ICE.

The ACE has a website where details can be obtained on the ACE's Scottish Group and committee. The website also provides information about the services offered by the ACE, amendments to ACE conditions of engagement, press releases and the ACE monthly journal.

The second trade association is the Civil Engineering Contractors Association which is a trade association for civil engineering contractors.

### 1.3.5 Specialist consultants

In large building projects employers often employ specialist consultants to advise on specific areas, for example, mechanical and electrical works or heating and ventilation works. Architects and engineers may be employed in a specific limited role to advise on an area in which they hold themselves out as specialists.

### 1.3.6 Project manager

Depending upon the size of the building contract and the method of procurement, the employer may decide to engage a project manager. A project manager is a particular feature of a management contract. The project manager's role tends to be organisational but it is important to define his exact relationship with the other members of the team. It should be noted that contractors can employ a person to organise the implementation of the contract and this individual can sometimes also be termed the project manager.

### 1.3.7 Clerk of works

A clerk of works is the individual on site who is generally employed by either the employer or the architect to oversee the execution of the works and to report back to the employer or architect.

### 1.3.8 Contractor

Once employers have decided upon the nature and extent of the work which they wish carried out (possibly with the assistance of the architect and quantity surveyor), they will usually invite one or more contractors to tender for the work. The term 'contractor' is used throughout this book as representing the individual, firm or company which is responsible for carrying out the building works for the benefit of the employer. It is adopted for the reason that it is the term used by the SBCC in its standard form building contracts and associated documentation. However other terms are to be found in general use such as 'the builder' or, in some instances, 'main contractor' to distinguish the contractor from sub-contractors. If the works to be carried out are minor works, such as joinery or electrical works then the more appropriate term for use may simply be 'the joiner' or 'the electrician'. The obligations of contractors under a building contract are considered below in Chapter 4.

### 1.3.9 Sub-contractors

In practice, elements of the work are often executed not by the main contractor but by sub-contractors or even sub-sub-contractors. These may include:

- 'nominated sub-contractors' (chosen by the employer or his agent according to special contractual rules);
- 'domestic sub-contractors' (chosen by the main contractor normally with the employer's consent);
- 'works contractors' (under a management contract) or
- 'labour only sub-contractors'.

The position of sub-contractors under a building contract is considered below in Chapter 10.

In addition to the principal parties referred to above, it must be appreciated that other parties can have a role in a building project such as suppliers (nominated or domestic), insurers, funders, prospective tenants and purchasers of the building.

## 1.4 *Types of building contract*

Once the employer has assembled his team, he needs to enter into a contract for the execution of the works. The contract documents are generally prepared on the advice of the architect and possibly the quantity surveyor. The employer and the other party to the contract, generally the contractor, are free to negotiate and thereafter determine the terms upon which they will enter into a contract with each other. This is considered in more detail below in Chapter 2. However, in practice, the negotiating parties rarely have equal negotiating power and one party can often impose its terms upon the other with little, or even no, negotiation. The terms of the contract will depend upon the nature of the work to be carried out and either implicitly or explicitly allocate between the parties the risks involved in the project, such as the risks of delay, disruption and increased costs.

There are many different contractual arrangements which can be entered into. In smaller projects there may only be a quotation by the contractor which is accepted, with or without qualification, by the employer. The quotation may, or may not, have standard terms and conditions attached to it. The parties may negotiate over the incorporation of all, or part of, the standard terms and conditions into their contract. Projects of a larger nature have now become more complex since, in recent years, an increasing number of contractual arrangements have become more common, particularly with regard to the responsibility for design and management of the project.

The following terminology is in common use when considering the types of building contract or form of procurement to be adopted for a building project.

### Lump sum contracts

A lump sum contract is a contract in which the contractor agrees with the employer to carry out the building works for a pre-agreed price. The price is only subject to adjustment in certain limited circumstances such as variations and fluctuations in costs.

### Measurement contracts

In this type of contract the sum which the employer pays the contractor is determined by measuring the work done on completion of the project and by applying it to agreed rates or some other form of valuation.

### Cost reimbursement contracts

There are different types of such contracts such as cost contracts, cost plus contracts and prime cost contracts but the common feature is that the sum which the employer pays the contractor is not a pre-agreed sum but a sum calculated by reference to the actual cost of the works carried out, generally with the addition of an amount to cover profit and a management fee. This may be a predetermined percentage of the costs, a predetermined fixed fee, or a variable fee calculated according to a predetermined formula.

### Design and build contracts

In a design and build contract it is the contractor who plays the main role by fulfilling both design and construction functions. If the contractor also undertakes management obligations, the contract may be referred to as a design manage construct contract.

### Package deals

It has recently become increasingly common for contractors not only to design the building and construct it but also to take full responsibility for some or all of the duties of the architect, engineer or even surveyor as performed in traditional contracts. This is sometimes referred to as a package deal.

### Turnkey contracts

This term is sometimes used in the context of design and build contracts. The contractor may undertake to provide a building ready for use, perhaps taking responsibility not only for the construction of the building but also for the procurement of the site and possibly even furnishing. It has been suggested that the term is intended to indicate that on completion of the project the key can be turned and the building will be ready for use. The phrase can also be used to refer to specific contract conditions. The European International Contractors (EIC) published the EIC Turnkey Contract in May 1994. FIDIC has also published what is known as its Orange Book entitled 'Conditions of Contract for Design-Build and Turnkey'.

### Management contracts

This type of contract is more common on large projects. There are different types of management contracts but the main feature of them tends to be that there is a project manager who undertakes to co-ordinate and organise the work rather than to construct it directly himself. There are broadly two different approaches. The project manager may be engaged by the employer along the same lines as construction professionals and, for a fee, undertakes to control and manage the project. Although the project manager will probably supervise the tendering process, the obligation to construct (and perhaps also to design) the work is undertaken by a contractor who contracts directly with the employer. Alternatively, a management contractor may undertake contractual responsibility for construction of the works but subcontract the functions of co-ordination and management of the project to a works contractor.

### Fast tracking

This is a type of contract where construction is commenced before the design of the whole project is complete with the aim of quick completion.

### Joint ventures

This type of arrangement is now a very common method of procurement. A special purpose company or partnership is created by two or more parties (often a land owner and a developer) each contributing their respective assets, funds and/or skills with a view to procuring a construction project. The joint venture company/partnership will become the employer for the purposes of the building contract.

## 1.5 SBCC forms of building contract

In Scotland many building contracts are entered into on the SBCC standard forms. The constituent bodies of the SBCC are currently the Royal Incorporation of Architects in Scotland, the Scottish Building Employers Federation, the Royal Institution of Chartered Surveyors in Scotland, the

Convention of Scottish Local Authorities, the National Specialist Contractors Council, Scottish Casec, the Association of Consulting Engineers (Scottish Group), the Confederation of Business Industry, and the Association of Scottish Chambers of Commerce. The SBCC is itself a member of Joint Contracts Tribunal Ltd, a company limited by guarantee (JCT). The RICS published a survey entitled 'Survey of Building Contracts in Use During 1995' which found that 85% of all contracts considered were let on JCT standard forms.

The SBCC has produced standard forms since 1964. Over the years, the number of standard forms has increased and numerous revisions and amendments can make it difficult to identify the precise terms upon which parties have contracted. Matters are often further complicated by the attempts of employers and contractors to modify the provisions of the standard form contracts which can result in unforeseen consequences for both the employer and the contractor.

JCT carries out from time to time a revision of its Standard Conditions; it did so in 1980 and again in 1998. The current edition of the Conditions is known as 'JCT 98'. The Standard Conditions are available in a number of variations for use depending upon the different circumstances of the contract. The conditions are regularly amended and care must therefore be taken to ensure that the correct print is being referred to. SBCC systematically revises its contracts and publishes for each JCT document the equivalent Scottish Building Contract. The Scottish Building Contract, in various versions, came into use in April 1980 and since then has been revised on various occasions. In early 1999 the SBCC commenced the exercise of reprinting a number of its standard form contracts to reflect the move from the 1980 to the 1998 edition. At the time of going to press this exercise had not been completed; however, where reprinting is anticipated this is noted below.

There is a wide variety of standard forms of contract available and care must be taken to select the correct version. The SBCC produce a *Guide to the Scottish Building Contract Committee/Joint Contracts Tribunal documents applicable for use in Scotland* which identifies the documents published by the SBCC. The SBCC publications and the Standard Form Contracts can be purchased from the RIAS or the RICS (see ss. 1.3.2 and 1.3.3 above).

In the remainder of this book, in examining the various matters which are of relevance to Scottish building contracts, we will consider the appropriate provisions of JCT 98. Unless the contrary is indicated, references to JCT 98 are to the conditions of the Standard Form of Building Contract Private With Quantities 1998 Edition as amended by, and incorporated into, the SBCC Scottish Building Contracts (May 1999 Editions).

### 1.5.1 SBCC Main Contract Documentation

The following are the main contract forms currently produced by the SBCC.

**Scottish Building Contract – With Quantities (May 1999 Edition)**

This is intended for use when the design is provided by the employer, normally through his architect, and the works are defined in advance in bills of quantities. The contract incorporates:

(a) The conditions of either:

(1) the Standard Form of Building Contract: Local Authorities With Quantities 1998 Edition; or
(2) the Standard Form of Building Contract: Private With Quantities 1998 Edition;

and the JCT supplemental provisions known as the VAT Agreement and Annex 2 to the JCT conditions.

The choice of conditions is made by deleting the appropriate reference in clause 3.1 of the contract.

(b) The Scottish Supplement forming Appendix I thereto;

(c) The Abstract of Conditions forming Appendix II thereto.

**Scottish Building Contract – Without Quantities (May 1999 Edition)**

This is intended for use when the employer supplies the design but the works are not defined in bills of quantities. A schedule of rates may be used in conjunction with this contract and, because there are various ways of defining the works, the contract provides for various methods of calculating the contract sum. The contract incorporates:

(a) The conditions of either:

(1) the Standard Form of Building Contract: Local Authorities Without Quantities 1998 Edition; or
(2) the Standard Form of Building Contract: Private Without Quantities 1998 Edition;

and the JCT supplemental provisions known as the VAT Agreement and Annex 2 to the JCT conditions.

(b) The Scottish Supplement forming Appendix I thereto.

(c) The Abstract of Conditions forming Appendix II thereto.

**Scottish Building Contract – With Approximate Quantities (May 1999 Edition)**

This contract is for use when the works have been substantially designed but not completely detailed so that the quantities shown in the bills are approximate and subject to remeasurement. The contract incorporates:

(a) The conditions of either:

(1) the Standard Form of Building Contract: Local Authorities With Approximate Quantities 1998 Edition; or
(2) the Standard Form of Building Contract: Private With Approximate Quantities 1998 Edition;

and JCT supplemental provisions known as the VAT Agreement and Annex 2 to the JCT conditions.

(b) The Scottish Supplement forming Appendix I.

(c) The Abstract of Conditions forming Appendix II thereto.

**Scottish Building Contract – Sectional Completion Edition – With Quantities or Without Quantities or With Approximate Quantities (May 1999 Editions)**

These contracts are designed for use where the works are to be carried out in sections, each of which is to be defined in advance. Separate completion dates and defects liability periods are fixed for each section. The employer takes possession of each section on practical completion of that section whereas under the other contracts he deals normally with this only once the whole works are practically complete. The contract sum is correspondingly divided. Each section has its own liquidated damages for delay. However, there is only one final certificate in respect of all works. The contracts incorporate:

(a) The relevant conditions which are either the Standard Form of Building Contract Local Authorities or Private Edition With Quantities, Without Quantities or With Approximate Quantities (1998 Edition) and the VAT Agreement and Annex 2 to the JCT conditions.
(b) The Scottish Supplement forming Appendix I.
(c) The Abstract of Conditions forming Appendix II.

**Scottish Building Contract – Contractors Designed Portion (May 1999 Editions)**

These contracts can be specified to be With Quantities, Sectional Completion/With Quantities, Without Quantities, or Sectional Completion/

Without Quantities. They were all reprinted in May 1999. They are for use where the contractor is requested to complete the design of part of the works ('the Contractor's Designed Portion'). They follow a similar format to that referred to above.

**Scottish Building Contract – With Contractors Design or With Contractors Design Sectional Completion (May 1999 Editions)**

These contracts are for use when the employer has issued his requirements to the contractor who has submitted proposals which include all necessary design work acceptable to the employer and where an architect/contract administrator has not been appointed to carry out the functions ascribed to an architect/contract administrator. These contracts follow a slightly different format to those referred to above. The contracts incorporate:

(a) The Conditions of the Standard Form of Building Contract With Contractor's Design 1998 Edition as modified.
(b) The Scottish Supplement forming Appendix I thereto.
(c) The Abstract of Conditions forming Appendix II thereto.
(d) An Appendix III entitled 'Alternative Methods of Payment'.
(e) The employer's requirements, the contractor's proposals and the contract sum analysis described in Appendix IV thereto.

**Scottish Minor Works Contract May 1999 Edition and Scottish Minor Works Sub-Contract (April 1995 Revision)**

These contracts should be used when the works are fairly straightforward. They do not provide for a price adjustment to take account of material fluctuations and are therefore generally unsuitable for long-term projects. The SBCC recommends that they should not be used for contracts lasting more than one year. They should also not be used where specialist sub-contractors for design input are to be employed. There are no optional provisions permitting the use of bills of quantities. No standard conditions or JCT amendments are incorporated. It is the intention of the SBCC to issue a new edition of the Scottish Minor Works Sub-Contract in due course.

**Scottish Measured Term Contract for Maintenance and Minor Works (May 1999 Edition)**

This contract is to be used when the employer requires maintenance work to be executed on a regular basis. It incorporates the conditions of the JCT Standard Form of Measured Term Contract 1998 Edition and supplementary conditions; a statutory tax deduction scheme; the Scottish Supplement forming Appendix I thereto and the Appendix II thereto.

**Scottish Management Contract (May 1999 Edition)**

This contract incorporates

(a) The Conditions of the Standard Form of Management Contract 1998 Edition; the Annex to the conditions and the VAT Agreement.
(b) The Scottish Supplement forming Appendix I.
(c) The Abstract of Conditions forming parts 1 and 2 of Appendix II
(d) The first, second, third, fourth and fifth schedules thereto.

Reference should also be made to Practice Notes MC/1 and MC/2.

**Scottish Management Contract - Phased Completion Edition (May 1999 Edition)**

This is used where a management contractor is being employed and the work is to be done in predesigned phases with the employer taking possession of each phase on practical completion. The contract follows a similar format to the Scottish Management Contract above. Reference should be made to the SBCC 'Notes for Guidance on the Scottish Management Contract - Phased Completion Edition'.

**Works Contract/1/Scot (May 1999 Edition)**

This consists of an invitation, a tender and a works contract. The Contract incorporates JCT Works Contract Conditions (Works Contract/2) 1998 Edition and stated supplemental provisions. The SBCC publish 'Notes for Guidance on Works Contract/1/Scot' (revised January 1994).

**Scottish Employer/Works Contractor Agreement Works Contract/3/Scot (May 1999 Editions)**

This contract should be read in conjunction with the SBCC 'Notes for Guidance on the Scottish Employer/Works Contractor/3/Scot (July 1991)'.

**Scottish Building Contract - Prime Cost Contract (July 1997 Revision)**

This incorporates the conditions of JCT Standard Form of Prime Cost Contract 1992 Edition together with appropriate amendments.

### 1.5.2 SBCC Domestic Sub-Contract Documentation

In September 1997 the SBCC produced a suite of domestic sub-contract documentation. In August 1998 revised versions of the sub-contract documentation were produced to take into account the provisions of the 1996 Act and certain associated difficulties that arose as a consequence of it. It is the intention of the SBCC to issue a new suite of sub-contract documents.

The SBCC has published Guidance Notes for the purposes of the Scottish Domestic Sub-Contract (September 1997 Edition - August 1998 Revision) to which reference should be made when using this suite of documents.

The SBCC domestic sub-contract documentation consists of the following.

#### Tender DOM/T/Scot/Part 1 (September 1997 Edition - August 1998 Revision)

An invitation to tender is issued by the main contractor to prospective domestic sub-contractors to be selected under clause 19 of JCT 80. The main contractor inserts all necessary information regarding both the main contract and the domestic sub-contract. The domestic sub-contractor's tender will be based thereon and it is therefore important that Part 1 is completed with all necessary information and signed by and on behalf of the main contractor.

#### Tender Dom/T/Scot/Part 2 (September 1997 Edition - August 1998 Revision)

The domestic sub-contractor acknowledges that he has received invitation to tender Dom/T/Scot/Part 1 and completes Part 2 with all necessary information. He thereafter signs it before returning it to the main contractor for consideration.

It should be noted that if a tender from a proposed domestic sub-contractor is not capable of being accepted without qualification then the main contractor must decide either to reject it or to discuss the qualifications with the proposed domestic sub-contractor. If agreement is not reached and the main contractor thereafter issues an acceptance (which will be on the basis of DOM/A/Scot) which does not meet the tender contained in Tender/DOM/T/Scot/Part 2, in law his acceptance is a counter-offer which the domestic sub-contractor is free either to accept or reject. In other words, until the domestic sub-contractor makes a decision either expressly or impliedly through his actings, no concluded domestic sub-contract will exist or, at the very least, there will be an uncertainty as to whether or not there is a concluded domestic sub-contract and, if so, on what terms. Until the position is clarified by discussion and negotiation (and among other things Appendix I is completed as agreed), the contractor is not in a position to secure the domestic sub-contractor's signature to Dom/A/Scot upon a reliable basis.

**Sub-Contract DOM/A/Scot (September 1997 Edition - August 1998 Revision)**

This is the Scottish domestic sub-contract and incorporates DOM/C/Scot by reference.

**DOM/C/Scot (1997 Edition - August 1998 Revision)**

This contains the standard conditions of domestic sub-contract for use in Scotland. These conditions are only suitable for use with the Domestic Sub-Contract Agreement (DOM/A/Scot) and where the main contract is on the Scottish Building Contract (post July 1997 revisions).

### 1.5.3 SBCC Nominated Sub-Contract Documentation

SBCC produce nominated sub-contract documentation which adopts, for use in Scotland, the nominated sub-contract documentation produced by JCT.

This suite of documents consists of the following.

**Tender NSC/T/Scot/Part 1 (May 1999 Edition)**

Standard Form of Nominated Sub-Contract Tender for use in Scotland. Part 1 - Invitation to Tender to a Sub-Contractor.

**Tender NSC/T/Scot/Part 2 (May 1999 Edition)**

Standard Form of Nominated Sub-Contract Tender for use in Scotland. Part 2 - Tender by a Sub-Contractor.

**Tender NSC/T/Scot/Part 3 (May 1999 Edition)**

Standard Form of Nominated Sub-Contract Tender for use in Scotland. Part 3 - Particular Conditions.

**Nomination NSC/N/Scot (May 1999 Edition)**

Standard Form of Nomination Instructions for use in Scotland.

**Agreement NSC/W/Scot (May 1999 Edition)**

Standard Form of Employer/Nominated Sub-Contractor Agreement.

**Sub-Contract NSC/A/Scot (May 1999 Edition)**

This Scottish Building Sub-Contract incorporates by reference Conditions NSC/C which are the Standard Conditions of Nominated Sub-Contract 1998, as amended or modified.

**Tender NSC/T/(PCC) Scot/Part 1 (September 1995 Edition)**

Standard Form of Nominated Sub-Contract tender for use in Scotland where the Scottish Prime Cost Contract is used. Part 1 – Invitation to Tender to a Sub-Contractor.

**Tender NSC/T/(PCC) Scot/Part 2 (September 1995 Edition)**

Standard Form of Nominated Sub-Contract tender for use in Scotland where the Scottish Prime Cost Contract is used. Part 2 – Tender by a Sub-Contractor.

**Tender NSC/T/(PCC) Scot/Part 3 (September 1995 Edition)**

Standard Form of Nominated Sub-Contract tender for use in Scotland where the Scottish Prime Cost Contract is used. Part 3 – Particular Conditions.

**Nomination NSC/N/(PCC) Scot (September 1995 Edition)**

Standard Form of Nomination Instruction for use in Scotland where the Scottish Prime Cost Contract is used.

**Agreement NSC/W/(PCC) Scot (September 1995 Edition)**

Standard Form of Employer/Nominated Sub-Contractor Agreement for use in Scotland where the Scottish Prime Cost Contract is used.

**Sub-Contract NSC/A/(PCC)/Scot (September 1995 Edition)**

This is the appropriate Scottish Building Sub-Contract where the Scottish Prime Cost Contract is used.

The Prime Cost Nomination suite of documents has not been revised in light of the 1996 Act, nor have new editions been printed in 1999.

### 1.5.4 SBCC Standard Forms for Nominated Suppliers

SBCC produce two documents in relation to this, namely:

- Standard Form of Tender for use in Scotland by a Nominated Supplier (January 1992 Revision)
- Schedule 3: Warranty Agreement for use in Scotland by a Nominated Supplier (January 1992 Revision)

Schedules 1 and 2 are incorporated in the Standard Form of Tender.

### 1.5.5 SBCC Standard Forms of Contracts of Purchase

These documents are required because of differences in the laws of Scotland and England relative to ownership and purchase of off-site materials. Three separate documents are produced by SBCC, namely:

- Contract of Purchase from the Contractor (Revised July 1991)
- Contract of Purchase from the Sub-Contractor (Revised July 1991)
- Contract of Purchase from a Works Contractor (May 1999 Edition)

### 1.5.6 SBCC Standard Forms of Collateral Warranties

Three separate forms of collateral warranty are currently produced by SBCC, namely:

- Collateral Warranty by a Main Contractor – MCWa/F/Scot (Funder) May 1999 Edition
- Collateral Warranty by a Main Contractor – MCWa/P & T/Scot (Purchaser) May 1999 Edition
- Standard Form of Employer/Sub-Contractor Warranty Agreement (Revision 6.11.98)

In addition to the SBCC and JCT a number of other bodies within the construction industry produce their own standard form contracts and associated documentation. An examination of these is beyond the scope of this book but readers should be aware that such documentation exists. Amongst others, such documentation is produced by ICE, Central Government, FIDIC, the Institution of Mechanical Engineers, Institution of Electrical Engineers, the Association of Consulting Engineers and, lastly, the Institute of Chemical Engineers. The use of documentation produced by other bodies may well be appropriate, depending upon the nature of the project in question.

# Chapter 2
# Entering into a Building Contract

## 2.1 Introduction

The majority of people do not realise the frequency with which they enter into a contract whilst going about their everyday business. For example, purchasing a train ticket constitutes the formation of a contract between the railway company and the passenger. Few people recognise that this is a formal legal arrangement which imposes rights and obligations on both the passenger and the railway company. In reality, contractual relationships of one nature or another hold the very fabric of the commercial world together, including the construction industry. Without the certainty that a contract provides, the resulting chaos would inevitably render the conduct of business, in any meaningful sense of the term, impossible.

## 2.2 Essentials of written and oral contracts

### 2.2.1 Agreement

A contract is essentially an agreement, expressed either in writing or verbally, between a number of parties (not necessarily restricted to two) regarding the same subject matter. The law relating to the formation of contracts is of general application notwithstanding the diversity of subject matters which may constitute the agreement between the parties. In this regard there is little distinction between, for example, a contract for the sale of goods and a building contract. The essentials of formation for both are identical.

A contract is formed when the parties to it reach agreement as to the essential elements of the transaction. There must be what is termed *consensus ad idem*. The literal meaning of *consensus ad idem* is 'agreement to the same thing' and it has also been expressed as a 'meeting of the minds'. There is no need for consensus between the parties in relation to every detail of the transaction – the test is an objective one. It was held in the case of *Muirhead & Turnbull* v. *Dickson* (1905) that,

> 'commercial contracts cannot be arranged by what people think in their inmost minds. Commercial contracts are made according to what people say ...'.

Care should be taken to distinguish between the courts' willingness to determine whether a contract exists and their reluctance to rewrite the agreement between the parties, see for example *Houldsworth* v. *Gordon Cumming* (1910).

Even where the parties to a contract do not think that they have reached agreement, there are many instances of the courts reaching the view that they have. In *Uniroyal Ltd* v. *Miller & Co Ltd* (1985) it was held that, in establishing whether parties had entered into a contract, the fundamental principle to be applied is to consider whether or not, and when, there has been *consensus ad idem* between parties. In this case there was no agreement regarding the price of goods to be supplied under the contract and it was held that that lack of consensus was fatal. The court held that there was no contract between the parties as, in this particular instance, price was a fundamental and essential part of the contract.

Performance of a purported contract in the mere belief by the parties to it that it was binding and where none of the essentials had been agreed will generally not be sufficient to enable the courts to conclude that *consensus ad idem* has been reached. In *Mathieson Gee (Ayrshire) Ltd* v. *Quigley* (1952) it was held that it is not enough for the parties to agree that there was a concluded contract where there was none due to the lack of agreement on essentials.

Even where there appears, on the face of it, to be no agreement there may still be a concluded contract as a result of the actions of the parties. In *Roofcare Ltd* v. *Gillies* (1984) the pursuers submitted a tender to carry out repairs to the roof of the defender's property. Their offer was made subject to the condition that the quotation was 'subject to the undernoted terms and conditions and no alterations, exclusions, additions, or qualifications to the quotation and specification will be made unless confirmed in writing by Roofcare'. The defender accepted the quotation, confirming that the pursuers should proceed with the repair work to the roof 'making same wind and watertight'. The pursuers did not reply to this qualification. The defender then allowed the work to proceed and the pursuers sued the defender for payment. The Sheriff granted decree against the defender who appealed on the basis that there was no *consensus ad idem* due to the wind and watertight qualification not being accepted by the pursuers, in effect there was no contract. The Sheriff Principal held that there had indeed been *consensus ad idem*, the pursuers had presented an offer that was the only basis upon which they would carry out the contract unless otherwise agreed in writing. The defender, who knew of that condition, added his qualification knowing that if it was to be accepted the pursuers would do so in writing. The pursuers did not accept the condition in writing. Despite this, the defender allowed the work to proceed. By his actions the defender was held to have accepted that the wind and watertight qualification did not apply and that there was a contract between him and the pursuers.

### 2.2.2 Offer and acceptance

The normal method of reaching an agreement will be through offer and acceptance, a simple analogy being the sale of a house by conclusion of missives. Using this analogy, when the offer to purchase the house is accepted by the seller without qualification the contract is concluded. As we have already seen, it is essential that there is *consensus ad idem* between the parties, that is, the acceptance must meet the offer.

An offer must be communicated to the party to whom it is made, see *Thomson* v. *James* (1855). It is thought that where an offer is communicated by a third party (unless an authorised agent of the offerer, for example, a solicitor) it cannot be accepted. Only where the offer is communicated by the party making it can it be accepted.

A simple, unconditional offer may be revoked at any time before acceptance, see *Thomson*. The revocation must be communicated to the recipient of the offer before it has any effect. Thus offerers may change their mind at any time prior to acceptance.

Where a time limit for acceptance is specified within the offer and no acceptance is received within that time, the offer will fall unless the offerer extends the time limit for acceptance. In *Thomson* it was held that an offer, pure and unconditional, puts it in the power of the party to whom it was addressed to accept the offer, until by the lapse of reasonable time he has lost the right. What is a reasonable time will depend on the facts and circumstances of each case.

A simple offer made without limit of time may lapse where there is a material change of circumstances after the offer has been made. In *McRae* v. *Edinburgh Street Tramways Co* (1885) it was held by Lord President Inglis that the change of circumstances must render the offer 'unsuitable and absurd' before it will lapse.

Where an offer is made and has not lapsed due to any of the above factors the contract will be concluded, provided there is agreement between the parties, when the offer is accepted. Acceptance can be express or it can be implied from the actions of the recipient of the offer. Again, it is essential that the acceptance is communicated to the offerer.

There is a general rule in Scots law that silence by the recipient of the offer does not imply acceptance of the offer. In *Wylie & Lochhead* v. *McElroy & Sons* (1873) it was held that the contention by the pursuers that the offerees' silence inferred acceptance was a most unreasonable one. Actings on the part of the offeree may be sufficient to infer that they have accepted the offer. In *Gordon Adams & Partners* v. *Jessop* (1987) the defender instructed the pursuers to place his property on a list of properties for sale. The pursuers, after inspection of the premises, wrote to the defender stating that the property was placed with them on a 'sole agency' basis. The defender's solicitor wrote to the pursuers stating that while the pursuers were instructed to place the property on the list, they were not appointed as sole agents. The pursuers responded that they would not accept a property unless it was on a sole agency basis. The defender did not respond to that but allowed the pursuers to continue to place the property on their list. It was held that a contract existed between the parties. The defender, in the full knowledge that the pursuers were insisting that they were sole agents, allowed the pursuers to place the property on their list. In the light of the defender's actions the pursuers' belief that there was a contract between the parties was a reasonable one, induced by the defender's behaviour.

Acceptance can be verbal, written or implied from the conduct of the parties. Above all, the acceptance must meet the offer. An acceptance which does not accept all of the parts to the offer or which tries to incorporate conditions or qualifications into the offer is not an acceptance at all but a counter-offer, see *Wolf & Wolf* v. *Forfar Potato Co* (1984). In general, the effect of a counter-offer is to refuse the original offer, which will then fall and can no longer be accepted. Where a counter-offer is accepted unconditionally by the original offerer then the contract will be concluded.

Acceptance of an offer must be communicated to the offerer before the contract is concluded. There are exceptions to this general rule, for example, where the contract is concluded as a result of the actions of the parties and where acceptance is made by post. The offer may stipulate the method of acceptance, for example, by post, telex, fax or telephone. Where the method of acceptance is stipulated communication of the acceptance must be made by that method or it will be invalid. Where no method of acceptance is stipulated, the acceptance is valid provided it is made in a competent manner. Particular rules apply regarding postal communications.

### 2.2.3 The postal acceptance rule

It is necessary to consider the impact of what is known as the postal rule at this stage. Again, the case of *Thomson* v. *James* (1855) is authoritative. In that case the offer was posted to the offeree. The offeree posted his acceptance and, on the same day, the offerer posted a letter withdrawing the offer. Both letters arrived at their respective destinations on the same day. The question for the court was which letter took effect first - was there a concluded contract or did the letter withdrawing the offer take effect before the letter accepting the offer? Obviously, if the retraction was effective first, then the offer no longer existed and could not be accepted.

It was held by the court that the acceptance was effective and that, therefore, there was a concluded bargain between the parties which could not be affected by the letter of revocation. This conclusion was reached on the basis that an acceptance is effective when posted, that is, the physical act of posting it means that the acceptance has taken effect, whereas a letter revoking an offer is not effective until it actually becomes known to the offeree.

Where the offer specifies a time limit within which it must be accepted, acceptance will be effective provided the acceptance is posted within the time limit. It is of no consequence to establishing whether there is a concluded contract if the acceptance is not actually received until a few days after the time limit expires provided it is posted before the time limit expires, see *Jacobsen, Sons & Co* v. *Underwood & Sons* (1894).

Where acceptance is made by telex transmission it has been held in the English case of *Brinkibon Ltd* v. *Stahag Stahl (1983)* that the postal rule does not apply – telex is a method of instantaneous communication and is therefore treated in the same way as an oral communication. A telex acceptance was held to be effective when printed out at the offerer's end. It is likely that facsimile transmissions would be treated in exactly the same way. The position regarding communication by e-mail is also likely to be as that with telex transmission. The Scottish Building Contract allows intimation of documentation in an adjudication by e-mail, see clause 41A.5.2. This should be contrasted with the determination provisions of clause 27.1, which are more traditional.

### 2.2.4 Battle of the forms

It is common within the construction industry for offers to be made subject to the offerer's standard conditions of contract (frequently printed on the reverse side of the offer or appended thereto). Difficulties arise where the offeree accepts the offer subject to the qualification that the offeree's standard conditions will apply. In the ordinary course of events this would undoubtedly constitute a counter-offer requiring the offerer's acceptance. Where, however, work is commenced prior to the counter-offer being accepted, a question arises as to whether there was, in fact, a contract and if so on whose terms. Whilst it has often been said of this scenario that the person firing the last shot will be successful, see *Butler Machine Tool Co Ltd* v. *Ex-Cell-O Corporation* (1979), it has also been commented that it may be more helpful to look at the documents as a whole to determine whether the parties have reached agreement on essential points, notwithstanding differences between the forms, see also *Butler Machine Tool Co Ltd*.

## 2.3 *Capacity to contract*

Special rules apply to certain categories of persons as to their capacity to enter into contracts. The main categories are as follows.

### 2.3.1 Young persons

The Age of Legal Capacity (Scotland) Act 1991 makes a distinction between two groups of young people, namely, those under the age of sixteen and those aged between sixteen and eighteen. Generally, a person under the age of sixteen has no legal capacity to enter into any transaction (s. 1(1)(a)). A transaction is defined to include contracts and unilateral promises (s. 9). Persons under the age of sixteen do, however, have limited ability to enter into contracts, that is, they have capacity to enter into contracts of a kind which are commonly entered into by persons of their age and circumstances and on terms which are not unreasonable (s. 2(1)). Any contract purported to be entered into by a person under the age of sixteen and who does not have legal capacity to contract will be void (s. 2(5)). The rights of parties to the void contract will be governed by the principle of unjust enrichment.

A person over the age of sixteen has legal capacity to enter into contracts. However, persons who enter into contracts between the ages of sixteen and eighteen can apply to the court to set aside the contract if it is shown to be prejudicial. Such an application must be made to the court before the person concerned attains the age of twenty-one (s. 3). A prejudicial transaction is one which

- an adult exercising reasonable prudence would not have entered into in the circumstances of the applicant at the time of entering into the transaction, and
- has caused or is likely to cause substantial prejudice to the applicant (s. 3(2)).

If the court makes an order setting aside the transaction then the rights of the parties will again be governed by the principle of unjust enrichment.

There are certain transactions specified in s. 3 that cannot be set aside on the grounds of prejudice. Such transactions include:

- a transaction in the course of the applicant's trade, business or profession (s. 3(3)(f));
- a transaction into which any other party was induced to enter by virtue of any fraudulent misrepresentation by the applicant as to age or other material fact (s. 3(3)(g)); or
- a transaction ratified by the applicant after they attained the age of eighteen years and in the knowledge that it could be the subject of an application to the court under s. 3 to set it aside (s. 3(3)(h)).

An application under s. 3 may not be made where the transaction has been ratified by the court under s. 4 (s. 3(3)(I)). A transaction can ratified by the court where a person between the age of sixteen and eighteen has entered into a transaction and all parties to the proposed transaction make a joint application to the court to have the transaction ratified (s. 4(1)).

### 2.3.2 Insanity

It is a general principle of Scots law that an insane person has no power to contract and any contracts which such a person purports to enter into are void, see for example *Gall* v. *Bird* (1855). In addition, such contracts are generally void even although the other party may not have known that he was dealing with a person of unsound mind at the time that the contract was entered into, see *Loudon* v. *Elder's Curator* (1923). However, continuing contracts in which a party has entered into whilst they were sane are not necessarily rendered void by that party's subsequent insanity.

### 2.3.3 Aliens

It is the position in Scots Law that, conforming to the Rome Convention, a contract made during a period of residence in Scotland cannot be set aside on the ground that one of the parties was an alien who lacked contractual capacity under his or her own legal system unless it is proved that the other party knew of the incapacity or was negligently unaware of it. This principle is incorporated into the Scottish legal system by s. 2 of the Contracts (Applicable Law) Act 1990.

### 2.3.4 Corporate bodies

A corporate body is a distinct legal entity which is entirely separate from the members of the corporation. A corporate body can enter into contracts and can sue and be sued. Corporate bodies will contract through their agents. The agent must have express or ostensible authority to bind the corporation to the contract he purports to make. Directors of companies have the power and authority to bind the company in transactions.

The doctrine of ultra vires requires consideration at this point. 'Ultra vires' means, quite literally, beyond its powers. A corporate body created by statute, or exercising statutory powers, cannot enter into any contract or dispose of its funds in any way which is not authorised by the statute or reasonably incidental to the powers conferred. Where the entity one is dealing with is a company incorporated under the Companies Acts the position regarding ultra vires has been simplified. Where a company deals with a person in good faith, the power of the directors to bind the company, or to authorise others to do so, is deemed to be free of any limitation under the company's memorandum and articles of association, see s. 35A of the Companies Act 1985. A person is not to be regarded as acting in bad faith by reason only of their knowing that an act is beyond the powers of directors under the memorandum and articles of association of the company. In addition, a person is presumed to have acted in good faith unless the contrary is proved, see s. 35A(2) of the Companies Act 1985. Neither is a party to a transaction with a company bound to enquire as to whether the transaction is permitted by the company's memorandum or as to any limitation on the powers of the board of directors to bind the company or to authorise others to do so, see s. 35B of the Companies Act 1985.

## 2.4 *Procurement*

As in many other areas of law, the European Union has had a significant impact in relation to tendering and procurement, especially with regard to public bodies, including local authorities, and the utility companies. The following intends to provide a brief overview of the present legislation affecting procurement and tendering. The principal European legislation has been given effect to in UK law by way of the Public Works Contracts Regulations 1991 (as amended), the Public Supply Contracts Regulations 1991, the Public Services Contracts Regulations 1993 and the Utilities Supply and Works Contracts Regulations 1992.

The regulations are intended to further the notion of a single market within the fields of procurement and tendering by ensuring that public contracts are awarded on the basis of objective criteria and not local preference. The regulations are also intended to add transparency to the tendering process thereby improving competition. A more detailed treatment of this area of law falls outwith the scope of this book.

At a national level successive government legislation has attempted to ensure that local authorities are more accountable for the decisions made in awarding works contracts. The introduction of compulsory competitive tendering has also attempted to fulfil policy objectives of introducing transparency to the tendering process and improving competition. Again, a detailed treatment of this area of law falls outside the scope of this book.

## 2.5 *Implied terms*

Implied terms are those which may be implied into a contract to reflect either the presumed though unexpressed intention of the parties or may be implied by statute or other rule of law irrespective of the intention of the parties, see *Morton* v. *Muir Brothers* (1907).

In relation to the former, which for the sake of convenience will be referred to as 'factual implication', a number of tests have been devised by the courts to determine whether the implication of terms may be permissible. It goes without saying that factual implication is heavily dependent upon the facts and circumstances of the case in question.

In *The Moorcock (1889),* Bowen LJ formulated a test for implying terms, which has become almost universally known as the 'business efficacy' test. He stated that:

> 'I believe if one were to take all the cases, and they are many, it will be found that in all of them the law is raising an implication from the presumed intention of the parties with the object of giving to the transaction such efficacy as both parties must have intended that at all events it should have.'

Another approach adopted by the courts is to apply what has become known as the 'officious bystander' test. The test derives its name from the judgment in the English case of *Shirlaw* v. *Southern Foundries* (1939). The test is essentially whether an 'officious bystander', a fictitious person who was privy to the discussions of the contracting parties, upon proposing the inclusion of a term would be 'testily suppressed with a common "Oh of course"' by the parties. Put more simply, is the term proposed by the 'officious bystander' so obvious that its intended inclusion goes without saying?

In either instance it is clear that the factual implication of a term must be reasonable in the circumstances, see *Morton*. However, not all terms that are reasonable may be capable of being implied – the courts to some extent require evidence of necessity as well, see *Liverpool City Council* v. *Irwin* (1977). Furthermore a term may also be implied on the basis of custom and usage, particularly in a district or trade or other context, see *Morton*. Implication under this head still requires evidence of necessity and to some extent overlaps with the officious bystander test, albeit it in a localised sense.

Implication arising by operation of a rule of law manifests itself in a number of ways. The implied duties that are incumbent upon a seller as to quality have long been elevated to statutory form and indeed still are, see the Supply of Goods and Services Act 1982. However, there are many instances where the present legislation applicable to sale is inapplicable. In such instances the courts have been willing to imply almost identical duties on the seller as those imposed under statute, simply because they are legal incidents to contracts of sale. However, in the absence of a precedent, implication on this basis will seldom arise, see *Scottish Power plc* v. *Kvaerner Construction (Regions) Ltd* (1998).

The intended scope of the following part of this section is restricted to a consideration of the implication of terms into contracts within the construction industry. A more detailed appraisal of implied terms lies outwith the scope of this book.

It would appear that the presence of an alternative remedy under a contract (albeit less attractive than the one sought under the implied term) will generally preclude the implication of an implied term. Thus where a contract provided for works to be carried out in phases and only one phase provided for an extension of time in the event of delay, the House of Lords refused to imply an extension of time clause into another phase where delay could have been dealt with under alternative provisions of that contract, see *Trollope & Colls Ltd* v. *North West Metropolitan Regional Hospital Board* (1973).

Over the years, the courts have become increasingly willing to imply terms which, in their most general form, have tended to require the employer and his agents (e.g. the architect or engineer) to fulfil their obligations timeously to allow a contractor to progress their works. Whilst each case will necessarily turn on its own facts, the following terms have been implied into main contracts by the courts:

- That the employer and its agent (in this case the engineer) were obliged to provide the contractor with all necessary details and instructions in sufficient time to enable the contractor to execute and complete the works in 'an economic and expeditious manner and/or in sufficient time to prevent the claimants being delayed in such execution and completion', see *Neodox Ltd* v. *Swinton and Pendlebury BC* (1958) and *J & J Fee Ltd* v. *The Express Lift Co Ltd* (1993).
- Not to hinder or prevent the contractors from carrying out their obligations in accordance with their contract or from executing the works in a

regular and orderly manner, see *London Borough of Merton* v. *Stanley Hugh Leach Ltd* (1982).

- The employer must take all reasonable steps to enable the contractor to discharge its obligations and to execute the works in an orderly and regular manner, including things which the architect is obliged to do to facilitate this, see *Mackay* v. *Dick and Stevenson* (1881) and *Lubenham Fidelities and Investments Co Ltd* v. *South Pembrokeshire DC and Another* (1986).
- The employer is obliged to ensure that the architect provides the contractor with full correct information concerning the contract work, see *Lubenham Fidelities Ltd.*

In *Neodox Ltd* v. *Swinton and Pendlebury BC* (1958) the court considered 'reasonable time' in the context of the provision of information. They said:

> 'What is reasonable time does not depend solely upon the convenience and financial interest of the claimant. No doubt it is to their interest to have every detail cut and dried on the day the contract is signed, but the contract does not contemplate that. It contemplates further details and instructions being provided, and the engineer is to have a time to provide them which is reasonable having regard to the point of view of him and his staff and the point of view of ... [the employer], as well as the point of view of the contractors.'

Similar duties to the foregoing would also appear to be incumbent upon a contractor when supplying necessary information and not hindering completion by a sub-contractor, see *J & J Fee Ltd* v. *The Express Lift Co Ltd* (1993).

Where the parties have clearly contemplated a risk, legal implication will not be sufficient to imply a term unless it satisfies the additional test of necessity, see *Martin Grant & Co Ltd* v. *Sir Lindsay Parkinson & Co Ltd* (1984). In addition implication will generally be precluded where the term seeks to impose liability on a party for matters over which they have no control, see *Ductform Ventilation (Fife) Ltd* v. *Andrews-Weatherfoil Ltd* (1995). Finally, the courts have been unwilling to imply a term where the implied terms sought were at variance with the express provisions of the contract, see *Martin Grant & Co Ltd.*

Implied terms in respect of quality are considered separately in Section 4.3.

## 2.6 Letters of intent

A letter of intent is a document that typically expresses an intention to enter into a legal relationship subject to the subsequent resolution of a condition. Such letters are intended to fall short of establishing a legal relationship and merely to provide comfort to the recipient. This area is less problematic under Scots law than English law. The English doctrine of consideration will generally ensure that a unilateral obligation (more commonly known as a promise) cannot be enforced. In general terms Scots law has no corollary with the doctrine of consideration thereby allowing a simple promise to be legally enforceable. The expression of a future intention to contract under a letter of intent is capable of being construed as a legally enforceable promise under Scots law. The Requirements of Writing Scotland Act 1995 s. 1 provides that a promise made in the course of business will be legally enforceable even where it is not expressed in writing. Thus a verbal expression of intention to contract, it is submitted, may be equally enforceable. Whether a letter of intent will fall short of establishing a legal relationship or constituting an enforceable promise will depend largely on the form of wording used in each case.

There is little Scottish case law on letters of intent. The case of *Uniroyal Ltd* v. *Miller & Co Ltd* (1985) would appear to be the only modern Scottish authority that deals explicitly with this issue. In this case Lord Allanbridge approved a reference to Gloag, in defining precisely the nature of a letter of intent, stating that:

> 'What is put forward as an acceptance may be read as a mere willingness to contract and of expectation that terms will be arranged.'

In many instances, where work has been carried out by one party pursuant to a letter of intent, the analysis of whether a contract has been formed is somewhat irrelevant; if there is no contract the party who has tendered performance will have a claim based on *quantum meruit* reasonable payment for work done. However, where a party who has received performance seeks damages for breach of contract, the issue is likely to be highly important. Where a letter of intent anticipates that, for example, a standard form of contract will be entered into by the parties at some future date and no such contract is subsequently entered into, the party who has received performance will be deprived of the protection which the terms of the standard form may have offered them. For example, in the absence of a contract stipulating the quality of the work to be performed, the party unable to avail themselves of the protection offered by the contract terms is left to argue that the *quantum meruit* claim should be reduced in light of the work performed.

English law is more developed in relation to letters of intent. The leading English authority on this point is *British Steel Corporation* v. *Cleveland Bridge & Engineering Co* (1984). Here, where a party commenced work on the basis of the words 'pending the preparation and issuing to you of the official form of sub-contract' contained in a letter of intent, it was held that it was 'very difficult to see how [the plaintiff], by starting work, bound themselves to any contractual performance'. Here, amongst other things, neither the price, the delivery dates, nor the applicable terms of contract had been agreed. It is worth noting that the omission of agreement as to price need not be fatal. The use of the word 'pending' was indicative of a state of preparation only.

Notwithstanding the analogy with promise, a letter of intent may also be construed as an offer capable of being accepted, depending on its terms, see *Uniroyal Ltd* v. *Miller & Co Ltd* (1985). In determining whether an offer has been made the terms of the letter itself are crucial.

## 2.7 Incorporation of terms by reference to another document

Generally, reference to a particular form of contract will be sufficient to incorporate its terms. Thus an agreement that a contractual relationship will be governed with reference to a particular form of contract will be sufficient to incorporate those terms into that contract, subject to the conditions referred to being readily identifiable or at least identifiable with reference to common industry knowledge, see *Modern Building Wales Ltd* v. *Limmer & Trinidad Co Ltd* (1975). Thus reference to a sub-contractor's order being 'in accordance with the appropriate form for nominated sub-contractors RIBA 1965 edition' was sufficient, after evidence had been led to show that, whilst a contract formally called 'RIBA 1965 edition' did not exist, the term was commonly used in the building trade to refer to the 'green form'.

The forgoing scenario envisages, however, that the terms of any conditions referred to will be suitable in the circumstances, for example, that a sub-contract relationship be governed by recognised sub-contract terms. What is more problematic and, indeed, a relatively common occurrence in the construction industry, is where party A attempts to impose the terms and conditions which they are subject to, for example, under a main contract, into a sub-contract which they have entered into with party B. This is commonly referred to as a 'back to back' arrangement.

This issue was considered by the Outer House of the Court of Session in *Parklea Ltd* v. *W & J R Watson Ltd* (1988). Here a sub-contract purported to incorporate the main contract conditions into the sub-contract that also contained other express terms. A dispute arose as to whether the arbitration clause in the main contract was applicable to the sub-contract. From this case it is submitted that a number of principles emerge which are of guidance in assessing whether such terms are capable of incorporation:

- The starting point must be to consider whether the parties have incorporated the whole of the main contract conditions; it is irrelevant that some (and not others) of the conditions would have fitted very neatly into the sub-contract conditions.
- Do the words incorporating the sub-contract conditions make clear that they are applicable to the exclusion of all other provisions? It was held that a reference to the main contract conditions 'solely regulating' the relationship between the parties was not indicative of an exclusion of all other conditions; the subsequent reference to the applicability of the main contract provisions being excluded where they conflicted with other express terms of the sub-contract mitigated against such a construction.
- Where the purportedly incorporated terms conflict or duplicate other express terms of the sub-contract or duplicate the terms of the main contract this will militate against the conclusion that the main contract terms will exclusively regulate the parties' contractual relationship.

The *Parklea Ltd* case follows a line of authority wherein the Scottish courts have been reluctant to apply the terms of an arbitration clause in similar circumstances. It is interesting to note that in *Parklea Ltd* it was a matter of agreement between the parties that only wholesale incorporation of the terms of the main contract would be sufficient to incorporate the arbitration clause. It would appear that the English courts might be prepared to adopt a broader approach. They have held where the main contractor's terms are not inconsistent with the sub-contractual relationship they could be incorporated, see *Brightside Kilpatrick Engineering Services* v. *Mitchell Construction (1973) Ltd* (1975), and, with particular regard to arbitration clauses, see also *Giffen (Electrical Contractors) Ltd* v. *Drake & Skull Engineering Ltd* (1993) and *Roche Products Ltd and Another* v. *Freeman Process Systems and Another (1996)*.

More in keeping with the approach of the Scottish courts is the Canadian case of *Smith & Montgomery* v. *Johnstone Brothers and Co Ltd* (1954) where it was held that where a contract made reference to the terms of another contract, and expressly incorporated a number of those terms, then only those expressly included will form part of that contract. Here the incorporation of an express term of a main contract, which made reference to the payment provisions in respect of nominated sub-contractors, and which was expressly incorporated into the sub-contract was held to be valid. The English courts have also addressed the issue as to whether the words 'shall be deemed to have notice of all the provisions of the main contract' are sufficient to incorporate those terms and answered in the negative, see *Jardine Engineering Corporation* v. *Shimizu Corporation* (1992).

## 2.8 *Signing a building contract*

The requirements of Scots law in relation to the signing of documents are now set out in the Requirements of Writing (Scotland) Act 1995 (which we will refer to as 'the 1995 Act').

Writing is *not* required for the constitution of a contract (s.1) except where the contract relates to the creation, transfer, variation or extinction of an interest in land. Although writing is not required for other forms of contract, the parties may execute their contract in such a way as to render the contract self-proving.

The 1995 Act distinguishes between a document which has been validly signed and a document which has self proving status. A validly signed document is one which has been subscribed by the grantor (s.2). Here extrinsic evidence is necessary to confirm the validity of the signatures. However, if the requirements of the 1995 Act regarding witnessing (which will be discussed in more detail later) have been followed then the signatures of the parties will be afforded self proving status. In effect this means by virtue of the means of execution the signatures of the parties are presumed and need not be proved (s.3).

Where a contract has schedules annexed to it, the schedules will be incorporated into the contract if they are referred to in the body of the contract and it is identified on the face of the schedules that they are the schedules referred to in the contract. If this is done there is no need for the schedules to be signed (s. 8(1)). It is only where a contract relates to interests in land that any schedule attached to the contract requires to be signed (s. 8(2)).

The precise requirements of subscription vary depending upon the designation of those signing the contract.

### 2.8.1 Individuals

If a party to a building contract is contracting as an individual (which includes a sole trader), then that person must subscribe the contract and have their signature witnessed by one witness. If this is done, subscription by that person will be self proving (s. 3(1)).

### 2.8.2 Partnerships

A contract will be validly executed on behalf of a partnership if it is signed by one partner or another authorised person (Schedule 2 para 2(1)(3)). The signatory can either sign his own name or the name of the firm (Schedule 2 para 3). The law regarding the power of a partner to bind a firm is set out in the Partnership Act 1890. The signature of the partner, or the authorised person, must be witnessed by one witness in order to make their subscription self proving.

### 2.8.3 Companies

A company will validly execute a contract if it is signed by one director, the company secretary or by a person authorised to sign the contract on the company's behalf (Schedule 2 para 3). Again, for the subscription to be self proving a single witness must witness it. The contract will also be self proving if it is signed by two directors, or a director and the secretary or by two authorised persons (Schedule 2 para 2(5)). In these circumstances there is no need for the signatures to be witnessed.

### 2.8.4 Local authorities

A contract will be validly executed by a local authority if it is signed by the proper officer, usually the chief executive, see Schedule 2 paras 4(1) and (3). A person purporting to sign as the proper officer is presumed to be the proper officer, see Schedule 2 para 4(2). For the subscription to be self proving, the contract must be subscribed by the proper officer on the local authority's behalf and either have the signature witnessed by one witness or have the contract sealed with the local authority's seal, see Schedule 2 para 4(5).

### 2.8.5 Witnesses

It is appropriate at this point to consider the law regarding witnesses. The 1995 Act reduced the requisite number of witnesses from two to one. If more than one signatory is signing at the same time one independent person can competently witness all signatures.

Witnesses must be independent with no direct interest in the contract. In addition, witnesses must be over the age of sixteen; be *compos mentis*; be able to write; and not be blind.

The witness must see the signatory sign the contract or, alternatively, the signatory can sign the contract outwith the presence of the witness and thereafter show their signature on the contract to the witness and acknowledge to the witness that the signature is in fact his. The witness must know the signatory but all that is required in that regard is a reliable introduction prior to signing or acknowledging, see *Brock* v. *Brock* (1908).

It is the practice for witnesses to sign opposite the signatory's signature and customary, although not strictly necessary, for witnesses to write the word 'witness' after their signature.

# Chapter 3
# Employers' Obligations

## *3.1 Introduction*

A building contract will usually set out, in express terms, the obligations owed by the employer to the contractor. Where these are not set out, certain terms will be implied due to the nature of the contract. Implied terms have been considered in Section 2.5. All of the obligations owed by the employer, be they express or implied, would appear to fall into two main categories, namely:

- an obligation of co-operation or the requirement to do certain things to put the contractor in a position of being able to carry out their own obligations under the contract; and
- an obligation to make payment for the work carried out by the contractor.

Although these will be referred to as employer's obligations, the employer commonly employs others to perform certain of these functions on their behalf, for example, the architect under JCT 98. The reference to the employer's obligations will therefore include the obligations which may be owed by other parties on the employer's behalf and it should be noted that breach by these parties will lead the employer to become liable for any losses suffered by the contractor as a result, see *Neodox Ltd* v. *Swinton and Pendlebury BC* (1958).

We will consider first the duties to do certain things necessary to enable the contractor to carry out his works. In *Mackay* v. *Dick and Stevenson* (1881), Lord Blackburn expanded on this obligation saying that it was

> 'a general rule that where in a written contract it appears that both parties have agreed that something shall be done, which cannot effectually be done unless both concur in doing it, the construction of the contract is that each agrees to do all that is necessary to be done on his part for the carrying out of that thing, though there may be no express words to that effect. What is the part of each must depend on circumstances.'

Whilst this might be described as the positive duty flowing from the employer's obligation to do everything necessary to enable the contractor to carry out their works, another term which has been frequently implied is the obligation not to do anything which will hinder the contractor from carrying out their obligations under the contract or from executing work in a regular and diligent manner, see *London Borough of Merton* v. *Stanley Hugh Leach Ltd* (1985). Examples of this general obligation to do all that is necessary to enable the contractor to carry out their works are;

- the obligation to give the contractor possession of the site;
- the obligation to administer the site; and
- the obligation to issue instructions and to provide information.

## 3.2 *Possession of the site*

### 3.2.1 General

In certain contracts it is an express term that the employer will give access to the contractor to enable him to carry out the works, see for example clause 23.1 of JCT 98. Where this is not expressly stated it will be implied as, in the majority of contracts, contractors cannot carry out their works unless they are actually on site, see *R* v. *Walter Cabott Construction Ltd* (1975). In referring to the requirement for the employer to give the contractor possession of the site, this is not possession in its legal sense but more a right of entry or control falling short of literal possession. The questions which then arise are when is the employer required to give the contractor possession what is the nature and extent of the possession which the employer requires to give the contractor (including consideration of whether or not the giving of possession implies that the contractor will have uninterrupted access to and possession of the site): and, finally, what is the duration of this obligation?

### 3.2.2 Time of possession

As stated above, the contract will often expressly state when the employer is required to give the contractor possession. Where the contract does not expressly provide for the date when possession must be given to the contractor, it will be implied that possession must be given within a reasonable time to enable the contractor to complete the works by any completion date which may exist, see *T & R Duncanson* v. *The Scottish County Investment Co Ltd* (1915). In the case of JCT 98, the actual date on which possession is to be given will be stated in the appendix to the contract. There is provision for postponement of this for a period not exceeding six weeks, or such lesser period as is stated in the appendix, see clause 23.1.2. An instruction given by the architect under clause 23.2 in regard to postponement is a relevant event by virtue of clause 25.4.5.1 that may entitle the contractor to an extension of time. It is also a matter affecting the regular progress of the works, by virtue of clause 26.2.5, which may entitle the contractor to loss and expense if the date of commencement of the contract is likely to have an overall effect on completion. Failure by the employer to issue an extension of time where they have delayed the commencement of the contractor's possession may result in the employer being unable to apply liquidated and ascertained damages where the contract has an expired completion date, see *Wells* v. *Army & Navy Co-operative Society Ltd* (1902).

### 3.2.3 Nature and extent of possession to be given

The contractor is normally entitled to possession of the whole site. That would appear to be what is meant by clause 23.1 of JCT 98 which refers to the contractor being given possession of the site on a specified date of possession. 'Site' is not a word that is defined by clause 1.3. In any case, it will be implied that the employer must make available the entire area that is necessary to enable the contractor to carry out their contract works. In some cases this has been held to extend beyond the actual area which will be occupied by the completed structure into other areas, for example, to provide working space and to enable the contractor to work efficiently and in accordance with generally accepted construction practices, see *R* v. *Walter Cabott Construction Ltd* (1975).

It is not an implied term, however, that the employer must provide work to the contractor in such a way as to enable them to carry out the work on an economic basis, see *Martin Grant & Co Ltd* v. *Sir Lindsay Parkinson & Co Ltd* (1984). A similar consideration arose in *Scottish Power plc* v. *Kvaerner Construction (Regions) Ltd* (1998). In that case the contract stipulated a period of 24 weeks as that allowed for the work. There was no guarantee of continuous working. The Lord Ordinary held that the defenders had power to interrupt the continuity of the period of 24 weeks. The reasoning behind that decision turns very much upon the provisions of the contract in question and it is submitted by the authors that the decision in *Scottish Power plc*, that where there is a specified contract period and no guarantee of continuous working the employer has the power to interrupt, is not one which should be followed. If this is an absolute right an absurd situation could arise with the employer being entitled to commence then stop the works at will.

The employer does not have any implied right to come on to site after possession has been given to the contractor. If they wish to retain the right to do so the contract should expressly provide for this, for example as clauses 11 and 12 of JCT 98 do in providing for the presence on site of the architect and clerk of works. However, it is submitted that possession should imply that the employer, or those employed by him, should have a right of reasonable access for the purposes of inspection, supervision and administration of the contract. Where nothing is said about possession, the contractor must be allowed use and possession of the site as required for the purposes of carrying out their works, see *Ductform Ventilation (Fife) Ltd* v. *Andrews – Weatherfoil Ltd* (1995).

However, after starting on site, the contractor may be denied undisturbed occupation for a variety of reasons, only some of which may be the responsibility of the employer. Where another contractor or supplier has been employed by the employer for certain aspects of the work not covered by the contractor's scope of works, the contractor, where he has had information regarding this other work, is under an obligation to permit the execution of such work. This will imply a right for the other contractors to enter onto the site to complete these other works, see for example, clause 29.1 of JCT 98. Where information in relation to such works that enables contractors to carry out and complete their own works has not been made available to the contractor, the employer's right to have access to, and to instruct other contractors to execute works on site, is subject to the consent of the contractor, see clause 29.2.

An employer will not be liable where there is unauthorised occupation by a third party such as picketers, at least not unless the employer has induced or condoned the obstruction, see *London Borough of Merton* v. *Stanley Hugh Leach Ltd* (1985). The employer does not warrant that access will not be prevented by a third party such as a picketer, see *LRE Engineering Services Ltd* v. *Otto Simon Carves Ltd* (1981). In the case of clause 2.2.2.1 of JCT 98, the incorporation of the standard method of measurement ('SMM 7') requires that any conditions relating to access should be stated in the bills of quantities. In certain circumstances, the employer may be liable to pay loss and expense to the contractor for failure to provide them with access to the site, see clauses 25.4.12 and 26.2.6. By virtue of clauses 4.1.1 and 13.1.2 of JCT 98, it remains open to the architect to vary the access available to the contractor, subject always to the contractor's reasonable right to object to this.

### 3.2.4 Duration of the obligation to give the contractor possession

The obligation subsists as long as the contract is running which will usually mean that the contractor ceases to have possession and the employer regains possession at practical completion. This is unless the contract provides for sectional completion and handover to the employer, for example, as provided for by clause 18 of JCT 98. Where there is no express provision stating when the contractor is to lose possession of the site, the contractor will be entitled to possession for so long as is necessary to allow them to perform their obligations under the contract, see *Castle Douglas and Dumfries Railway Company* v. *Lee, Son and Freeman* (1859). An intervening event may occur which allows the employer to take back possession of the site prior to practical completion, where for example they are entitled to determine the contract. In those circumstances, under clause 27.5.3 of JCT 98 (which is amended by the SBCC Scottish Building Contracts) the contractor may be obliged to remove or have removed from the site any temporary buildings, plants, tools, equipment, goods and materials. Determination is considered below in Section 8.4.

## 3.3 *Administration*

### 3.3.1 General

Employers are under an obligation to administer the site in such a way as to ensure that contractors can meet their obligations under the contract. Under this heading there will be considered the obligations incumbent on the employer under JCT 98 contracts to appoint an architect and to nominate sub-contractors where the contract calls for this, and the obligation not to interfere with the certifying process where a certifier, such as an architect, has been appointed. The obligation upon employers to use their best endeavours to ensure that the architect carries out his required functions, where it appears that he may be failing to do so, will also be considered under this heading.

### 3.3.2 Appointment of architect

The SBCC Scottish Building Contracts require the appointment of an architect (or contract administrator, as the case may be) together with a quantity surveyor if appropriate, see clause 4. The architect acts in all respects as the employer's agent. Normally, the architect will have been appointed prior to the contractor tendering for the contract. Failure to appoint an architect where the employer is contractually obliged to do so is a breach of contract by the employer, see *London Borough of Merton* v. *Stanley Hugh Leach Ltd* (1985). Where the contract calls for the appointment of an architect, it may well be that this is a condition precedent to the contractor's obligation to perform the work. Contractors may, however, be personally barred from insisting on the appointment of the architect, for example, if they proceed to work and the contract proceeds without the appointment of an architect.

If for any reason the architect becomes unable to act, the employer has a duty to appoint another architect. They should do so within a reasonable time (see clause 4 of the Scottish Building Contract) and their refusal to do so may amount to a repudiation of the contract entitling the contractor to rescind. Repudiation and recission are considered below in Sections 8.5 and 9.2. Whilst in most contracts the identity of the architect will be expressly stated it would be wise for the contract to be worded to refer to the appointment of the individual architect 'or such other person as may be nominated by the employer'. This is in order to avoid a situation arising where there may be confusion surrounding the existence of an obligation to appoint a successor should the appointment fail for any reason. In *Croudace Ltd* v. *London Borough of Lambeth* (1986) it was held that there had been a breach of contract on the part of the council where the architect employed by them on a contract, and who had been dealing with the contractor's claim for loss and expense, retired and the council delayed in appointing a successor. In that case, the architect named in the contract had a responsibility to ascertain the contractor's claims. There would appear to be an obligation on the employer to ensure that the successor to the original architect is reasonably competent to perform the job, see *London Borough of Merton*.

### 3.3.3 Nomination of sub-contractors and suppliers

An employer may decide to nominate a sub-contractor or supplier where, for example, they wish to ensure the quality of certain work that is to be performed, or the quality of certain materials that are to be supplied, or to avoid the price constraints which the contractor may be under. In these circumstances the sub-contractor/supplier is termed a nominated sub-contractor/supplier. JCT 98, as amended by the SBCC Scottish Building Contracts, sets out the obligations owed by the architect (as the employer's agent) in these circumstances, see clauses 35 and 36. There is an obligation on the employer to appoint nominated sub-contractors in reasonable time so as to allow contractors to carry out their obligations under the contract. The main contract should be operated in such a way that the terms of appointment of the nominated sub-contractor are agreed prior to the main contractor accepting the nomination.

By virtue of clause 35.5.1, the main contractor is entitled to make objections to proposed nominated sub-contractors. Such objections must be in writing at the earliest practicable moment, but in any case, not later than seven working days from the receipt of the architect's instruction nominating the particular sub-contractor. For example, a main contractor may object to a nomination where the proposed nominated sub-contractor has a later completion date than that which he is subject to under the main contract.

The late appointment of a nominated sub-contractor may, but does not necessarily, entitle the contractor to an extension of time and recovery of loss and expense provided that the appropriate application is made. If the employment of the first nominated sub-contractor is determined under clause 35.24, the architect is under an express obligation to appoint another nominated sub-contractor. This re-nomination should be done within a reasonable time, see *North West Metropolitan Regional Hospital Board* v. *T A Bickerton & Son Ltd* (1970). Again, if the re-nomination is delayed, the contractor may be, but is not automatically, entitled to an extension of time and recovery of loss and expense.

In the absence of an express provision requiring an employer to make a second nomination if the first nominated sub-contractor repudiates his sub-contract, it is submitted that there is no implied obligation on the employer to make a second nomination. The nominated sub-contractor will remain contractually bound to the main contractor, assuming that the repudiation happened after the main contractor had accepted the nomination of that particular sub-contractor. Accordingly, a second nomination would be inappropriate.

### 3.3.4 Obligation of non-interference

The architect's role is quasi arbitral in nature. This is considered below in Section 6.6. Whilst he is a professional person, he is not independent. He is an agent of the employer, see *Beaufort Developments (NI) Ltd* v. *Gilbert-Ash NI Ltd and Another* (1998). The employer is nevertheless under an implied obligation not to interfere with the operation of the certification process by the architect. Employers may be open to a claim for damages should they attempt to do so. Employers owe a duty to ensure that the architect discharges his obligations properly, see *London Borough of Merton* v. *Stanley Hugh Leach Ltd* (1985). They may also owe to the contractor a duty to replace an incompetent architect where they become aware that the architect is failing to perform his functions under the contract, or is taking into account things he ought not to, having regard to the contract, see *Panamena Europea Navigacion Compania Limitada* v. *Frederick Leyland & Co Ltd* (1947).

## 3.4 *Information and instructions*

JCT 98 provides that from time to time, as and when it may be necessary, the architect will issue such further drawings or details as are reasonably necessary either to explain and amplify the contract drawings or to enable the contractor to carry out and complete the works in accordance with the conditions, see clause 5.4. Should the contractor not receive information and instructions in the necessary due time where he has specifically applied in writing to the architect for the provision of such information (provided that his application was made at a time which, having regard to the completion date, was neither unreasonably distant from nor unreasonably close to the date on which it was necessary for him to have this information) then this is a relevant event which may entitle the contractor to an extension of time (by virtue of clause 25.4.6.2) and also a matter materially affecting regular progress of the works which may entitle the contractor to recover loss and expense, by virtue of clause 26.2.1.2. These obligations of the employer correspond with the obligation of the contractor, contained within clause 4.1.2, to comply with instructions issued by the architect within the stipulated time. The obligations of the contractor are considered below in Chapter 4.

In *Neodox Ltd* v. *Swinton and Pendlebury BC* (1958), it was held that what was a reasonable time for the provision of details and instructions necessary for the execution of the work did not depend solely on the convenience and financial interests of the contractor. The employer, through his agents (the engineer in this case), was to have a period of time to provide the information which was reasonable having regard to the point of view of himself and his staff, as well as that of the contractor. It was held in this case that there was an implied term that details and other instructions necessary for the execution of the works should be given by the employer's agent from time to time in the course of the contract and should be given within a time reasonable in all the circumstances.

Employers, through their architects, will therefore be in breach of contract for failure to give details and information in sufficient time to enable contractors to perform their obligations under the contract. This does not imply an obligation to provide information to contractors such that they can complete ahead of the contractually stipulated date, even if they have indicated that this is their intention, see *Glenlion Construction Ltd* v. *The Guinness Trust* (1987).

As regards requests from the contractor for information and instructions required by him, it has been held that a document setting out in diagrammatic form the planned programme for the work and indicating the days by which instructions, drawings, details and levels were required, which was issued by the contractor at the commencement of the work, could amount to a specific application for information in terms of the JCT 80. This was provided that the date specified for delivery of each set of instructions met the contractual requirement of not being unreasonably distant from nor unreasonably close to the relevant date, see *London Borough of Merton* v. *Stanley Hugh Leach Ltd* (1985).

Failure by the employer to provide the contractor with the drawings and necessary information to enable them to carry out their works may, depending on the importance of the work in question and after a reasonable request for the information by the contractor, constitute a repudiation of the contract entitling the contractor to rescind. Clause 28.2.2.1.2 specifically provides that such failure to give instructions and details can, in certain defined circumstances, be a specified event entitling a contractor to determine the contract. Determination is considered below in Section 8.4.

## 3.5 *Variations*

### 3.5.1 General

To instruct variations might be more appropriately described as a right the employer has, which imposes a corresponding obligation on the contractor to carry out the work so instructed, as well as an obligation on the employer to pay for the varied work. The nature of variations and the treatment of them under JCT 98 will be considered here.

In general, neither of the parties to a building contract has an implied right to vary the works on the basis that, having entered into a contract to carry out certain works for a specific sum of money, the parties are entitled to stand by this and do no more than that which they have contracted to do. In reality, however, the work which was originally specified may have to be modified for a variety of reasons such as unexpected physical ground conditions or other contingencies which parties were not able to identify with any degree of certainty at the outset. This is particularly so on a major building project. In addition the employer may wish to instruct the contractor to carry out certain extra works or, having discovered a quicker or easier way of doing something, to omit certain works that were originally included within the contract.

For these reasons, the contract, if in written form, will almost invariably entitle the employer to vary the works and a contractor will be under an obligation to carry out or omit works as either instructed or omitted by way of a variation instruction. This is subject to the contractor's right to be paid for the work and to receive an extension of time if the contract completion date is delayed as a result of carrying out the additional work instructed.

Essentially, a variation is a change to the scope of the contract works and is something that will lead to an adjustment of the contract price. Any work which the contractor is either expressly obliged to do in terms of the contract, or which is necessary by implication, falls within the contractual scope of the works and can never amount to a variation. It will therefore depend on the terms of the contract whether or not an instruction to carry out certain works will amount to a variation of the contract works for which the contractor should be entitled to additional payment.

For example, in a contract which can truly be said to be lump sum there is no obligation on the employer to pay for work by way of a variation even if the work is not described or shown on drawings or if the contractor incurs additional costs due to the impracticable nature of the design. If the contract obliges the contractor to achieve a particular result the contractor cannot claim as a variation a requirement to use more expensive materials if it becomes obvious that cheaper materials will not be appropriate. This is a risk the contractor takes in tendering a sum to achieve that end result. Likewise, if something is missing from the bills of quantities but, nevertheless, is necessary to achieve the end result, the requirement for the contractor to do that work will not amount to a variation. For example, in *Williams* v. *Fitzmaurice* (1858) a contractor was obliged to build a house which was to be ready by a specific date. The specification for the works did not include for any type of flooring and the contractor tried to state that he was entitled to extra payment for having to fix floorboards. However, as his contract was to achieve a particular result, namely, the completed house, the flooring was deemed to be included in the contract and this did not amount to a variation.

The contract will often provide for the employer having power to vary the contract works (usually through the architect) and in these situations, the contractor will be obliged to comply with the instructions of the architect, which instructions, in the case of JCT 98, should be in writing, see clause 4.3.1. Where employers have varied the contract works their obligation to provide detailed drawings in respect of the varied works will be the same as their obligation to do so in relation to the contract works; that is to provide such drawings and information within a reasonable time to allow the contractors to perform their obligations in terms of the contract.

In interpreting the scope of the contract, whether or not the bills of quantities form a contract document will be of great importance. In the past, bills of quantities were considered to be only an estimate of the works that were used as a guide for the contractor. They were not contract documents and were not to be taken as having contractual effect. If more materials were required than stated in the bills this did not amount to a variation of the contract works. Nowadays, bills of quantities may be incorporated as a means of pricing varied or omitted work that has been instructed by the architect. Where the bills of quantities are incorporated as a contract document, that is the contractor is to carry out the work as per the bills of quantities, contract drawings and specification, then an increase in the quantities will amount to a variation. This can be the case even if the contract is said to be lump sum, see *Patman & Fotheringham Ltd* v. *Pilditch* (1904).

Where the contract gives the employer or his architect power to order variations it will specify the extent of this power and will provide that variations can be issued at any time up to completion of the works, subject to appropriate adjustments being made to the completion date and the contract price, see clauses 25.4.5 and 30.6.2.11 respectively.

### 3.5.2 Variations under JCT 98

In JCT 98, the term 'variation' has a detailed definition, which is set out in clause 13.1. As used in JCT 98, the term variation has two separate meanings.

- First, it means the alteration or modification of the design, quality or quantity of the works. This includes the addition, omission or substitution of any work; the alteration of the kind or standard of any of the materials or goods to be used in the works; or the removal from the site of any work executed, or materials or goods brought onto site by the contractor for the purposes of the works (other than work, materials or goods which are not in accordance with the contract).
- Second, the term variation means the imposition by the employer of any obligations or restrictions in regard to four defined matters (to which we will return below) or the addition to, or alteration or omission of, any such obligations or restrictions imposed by the employer in the contract bills in relation to the defined matters.

The four defined matters are:

- access to the site or use of any specific parts of the site;
- limitations of working space;
- limitations of working hours; and
- the execution or completion of the work in any specific order.

There is specifically excluded from the definition the nomination of a subcontractor to supply and fix materials or goods or to execute work of which the measured quantities have been set out and priced by the contractor in the contract bills for supply and fixing or execution by the contractor.

The architect's powers in respect of variations are, largely, set out in clauses 13.2 and 13.3. The architect is expressly given the power to issue instructions requiring a variation and any instruction issued by him in this way is subject to the contractor's right of reasonable objection. The contractor's right of reasonable objection is contained within clause 4.1.1.

The architect also has the authority to issue instructions to expend provisional sums included in the contract bills or in a nominated sub-contract. The architect is empowered to sanction in writing any variations made by the contractor otherwise than pursuant to an instruction of the architect. Variations should be instructed or confirmed in writing by virtue of clause 4.3.1, save where they arise automatically, for example, where the contractor is working under different conditions as envisaged by clause 13.1.2, or where there has been an error in the contract documents or a departure from SMM 7 in accordance with clause 2.2.2.2.

The issuing by the architect of an instruction requiring a variation is a relevant event under clause 25.4.5 entitling the contractor to an extension of time, but only insofar as the works have in fact been delayed by the issue of the instruction and so long as the contractor has followed the requirements of clause 25. The contractor is also entitled to payment for the work carried out in complying with the architect's instruction and JCT 98 provides detailed provisions as to how such work should be valued. Those provisions are to be found in clauses 13.4 to 13.7.

The employer cannot vary the contract to the extent that it alters the fundamental nature of the contract works nor can he vary the contract by omitting large aspects of it and then employing another contractor to carry out the work. That will amount to a repudiation which, in turn, entitles the contractor to rescind and seek damages.

Although under JCT 98 there is a requirement for variation instructions to be in writing, this may be waived in certain circumstances. Examples of such circumstances are

- where the work is of such a different character or nature that it is said to be outside the terms of the contract and forms a separate contract;
- where the main contract is no longer operative;
- where the final certificate has been issued including a sum for the variations and there is no provision for review of this;
- where the arbiter is given the power to consider whether the work is or is not a variation and he decides that the work done was a variation; or
- where, for any other reason, the employer is personally barred from insisting, or has waived his right to dispute, that something was properly a variation where the instruction was not in writing.

## 3.6 Other obligations

### 3.6.1 Payment

The other main obligation owed by the employer to the contractor under a building contract, as outlined at the start of this chapter, is to make payment for the works executed under the contract. The obligation incumbent on the employer to make payment to the contractor is of such significance that it is considered separately in Chapter 7. In the context of JCT 98 payment is conditional upon the issue of certificates which are, therefore, also significant in the context of payment. Certification is considered below in Chapter 6.

### 3.6.2 Insurance and indemnity

In considering the obligations incumbent upon employers under building contracts, the issues of insurance and indemnity are also worthy of mention. At common law, there is no implied obligation incumbent upon an employer, to insure, however, frequently, the form of contract used by parties, such as JCT 98, will include such obligations. These are considered separately in Chapter 12 below.

### 3.6.3 Health and safety

As with all other employers in the employer/employee sense, as opposed to the building contract sense, employers owe a number of duties in respect of health and safety. Similar duties are incumbent upon contractors. In the area of building contracts such duties are, generally, more pertinent to contractors. These matters are considered below in Section 4.7.

# Chapter 4
# Contractors' Obligations

## 4.1 *Introduction*

As with the obligations of the employer, a building contract will ordinarily set out, in express terms, the obligations owed by the contractor to the employer. Similarly, where these are not expressed, certain terms will be implied into the parties' contract. The majority of the obligations we will consider in this chapter relate to the execution of the works, but it should be borne in mind that parties are free to contract in whatever manner they see fit. Accordingly, certain other types of obligation are routinely imposed upon contractors, for example, the obligation to take out and maintain insurance under JCT 98.

## 4.2 *Completing the works*

### 4.2.1 Common law

Where a contractor is engaged to carry out specified work, they have an obligation to carry out and complete that work. This carries with it the obligation to execute the work in a good and workmanlike manner using the skill and care to be expected of a builder of ordinary competence. This involves adopting methods which are in accordance with the regular practice in the building trade at the time, see *Morrison's Associated Companies Ltd* v. *James Rome & Sons Ltd* (1964). The exception to this would be if, in particular circumstances, there was an indication of an unusual or extraordinary risk in doing the work in the normal manner. In these circumstances, the contractor would be required to carry them out in a different way or else would run the risk of being found negligent, see *Morrison's Associated Companies Ltd.*

### 4.2.2 JCT 98 provisions

The JCT provisions expressly include the obligation to carry out and complete the works. This is to be found in clause 1 of the SBCC Scottish Building Contract and clauses 1.5 and 2.1 of the JCT 98 conditions. The obligation is to carry out and complete the works in accordance with the contract documents.

As the work proceeds, the contract allows the architect to issue instructions and the contractor has an obligation to comply with all relevant instructions issued to him in regard to any matter over which the architect has power to issue instructions under clause 4.1. The instruction may be one which requires a variation under clause 13.2.1. Variations are considered above in Section 3.5.

The contract contains a corresponding obligation on the employer to make payment for any varied work carried out. It provides, at clause 13.4.1.1, that all variations required by an instruction of the architect shall be valued by the quantity surveyor and sets out the basis for valuation at clause 13.5. The amount of the valuation is then used to adjust the contract sum, see clause 30.6.2. It is possible for a variation to omit work. In that case the value of the work omitted is calculated by reference to the rates and prices of the work as set out in the contract bills by virtue of clause 13.5.2. This value is then deducted from the contract sum in terms of clause 30.6.2.3. Payment is considered in detail below in Chapter 7.

A further matter arising from variations is the contractor's entitlement to additional time. Under clause 25.2.1.1 the contractor is obliged, if and when it becomes apparent that progress of the work is being or is likely to be delayed, to give written notice to the architect of the circumstances of this. The notice should identify any matter which is a relevant event. The contract contains a list of matters which are termed relevant events, one of which is the compliance with architect's instructions to vary the works, see clause 25.4.5.1. Upon an application by the contractor, if the architect considers completion of the work is likely to be delayed, then he grants an extension of time to the contractor and fixes a later date as the completion date all in terms of clause 25.3.1. Extensions of time are considered in more detail below in Chapter 5.

Finally, variations can also give rise to loss and expense being payable to the contractor. As soon as it becomes apparent to the contractor that regular progress of the work has been or is likely to be affected by one of the matters listed in the contract, the contractor can, in accordance with clause 26.1, make written application to the architect stating that he has incurred or is likely to incur direct loss and/or expense for which he would not be reimbursed by a payment under any other provision of the contract. One of the matters to which this applies is architect's instructions requiring a variation, see clause 26.2.7. If the architect considers that direct loss and/or expense has been incurred he ascertains (or instructs the quantity surveyor to ascertain) the amount of direct loss and/or expense sustained by the contractor. Any amount so ascertained falls to be added to the contract sum in accordance with clause 26.5.

The JCT 98 contract defines completion of the works as being when in the opinion of the architect practical completion of the works is achieved and the contractor has complied with the requirements of clause 6A.4, which relates to provision of information for the health and safety file that is required by the CDM Regulations. At this date the architect issues the certificate of practical completion under clause 17.1. This has been taken as meaning that the works have been completed for all practical purposes and the employer could take them over and use them for their intended purpose, see *Borders Regional Council* v. *J.Smart & Co (Contractors) Ltd* (1983). The issue of the certificate of practical completion triggers the release of the first half of the retention fund under clause 30.4.1.3.

The employer may take possession of the works once practical completion has been achieved by virtue of clause 23.3.1. This does not put an end to the contractors' obligations. They remain responsible for remedying any defects, shrinkages or other faults which may appear during the defects liability period and which are due to materials or workmanship not in accordance with the contract or to frost occurring before practical completion, see clause 17.2. This is more fully dealt with in Section 4.3 below.

Once the defects, shrinkages and other faults as specified in the schedule have been made good, the architect issues a certificate of completion of making good defects under clause 17.4. This triggers the release of the second half of the retention fund under clause 30.4.1.3.

## 4.3 *Quality of the work*

### 4.3.1 Workmanship

As far as workmanship is concerned, at common law, the contractor has an obligation to execute the work in a good and workmanlike manner using the skill and care to be expected of a builder of ordinary competence. This obligation subsists whilst the works are being carried out, it does not only arise at completion, see *Surrey Heath BC* v. *Lovell Construction Ltd and Another* (1998).

The obligation goes further than simply carrying out the work as it is contained on drawings or in other design information provided by the design team. In a case where the contractors were to provide all materials and perform all work shown on an architect's drawings, but where they were not supervised by any architect or engineer, it was found that the contractors had accepted that the employer was relying on their skill as contractors. There were defects in the plans provided to the contractor. The contractor ought to have recognised the defects and, since they were being relied on by the employer, had a duty to warn the employer of the defects in the plans and the difficulties which would arise if the plans were followed, see *Brunswick Construction Ltd* v. *Nowlan & Others* (1974). Where a contractor expressly undertakes to carry out work which will perform a certain duty or function in conformity with plans or specifications, a 'fitness for purpose' obligation, and it turns out that the works would not perform that function if the plans were followed, the contractor will be liable for the failure to perform even if the work was carried out in accordance with the plans and specifications.

The courts have been willing to imply a term into a contract requiring a contractor to warn of design defects as soon as the contractor came to believe that they existed. This was on the basis that if, on examining the drawings or as a result of experience on site, a contractor formed the opinion that in some respect the design would not work, or would not work satisfactorily, it would be absurd for them to carry on implementing it, see *Equitable Debenture Assets Corporation Ltd* v. *William Moss Group Ltd and Others* (1984) and *Victoria University of Manchester* v. *Hugh Wilson and Others* (1984). This does not oblige the contractor to carry out a critical examination of the drawings, bills and specifications looking for mistakes. The contractor's primary duty is to build not to scrutinise the design. The obligation to warn arises when, in the light of their general knowledge and practical experience, the contractor believes that an aspect of design is wrong, see *Victoria University of Manchester.*

Even though the contractor has to be satisfied that the design is satisfactory or suitable, the work still requires to be carried out in accordance with good building practice. A contractor cannot rely on drawings or designs produced by a surveyor to relieve them of this duty, see *Mackay* v. *Stitt* (1988). The contractor, if not satisfied with a design that is produced to him, has a duty to raise any queries with the design team. Even if the contractors are given such assurances they may, if still not satisfied, need to take precautions against failure of the design. If they do not do so then they may be found to have acted with less care than is to be expected of an ordinary competent builder, see *Edward Lindenberg* v. *Joe Canning and Others* (1992).

Where the contractor is not responsible for the design of a system or its integration into the works or for the selection of a proprietary system, there is no implied warranty by the contractor that the system will work, see *Greater Glasgow Health Board* v. *Keppie Henderson & Partners* (1989).

The view of what constitutes normal practice may alter depending upon the nature of the development. For example, where plumbing sub-contractors in a multi-storey flat development took the normal steps which would be taken to drain pipes of water in ordinary houses or small developments, they were found to be to blame for damage caused by burst pipes – the burst being caused by water which had not been drained from the pipes. The plumbing sub-contractors had failed to warn anyone that the pipes could not be completely drained. The normal practice of draining pipes, which resulted in some water remaining, did not apply in the face of the extent of the damage which might be expected in a multi-storey development were this to be followed, see *Holland & Hannen & Cubitts (Scotland) Ltd* v. *Alexander Macdougall & Co (Engineers) Ltd* (1968).

Where employers make known the purpose of the building and circumstances indicate that they are relying upon the contractor's skill and judgment to provide it, there is an implied term that the works will be fit for the purpose for which they were intended. In a case concerning the collapse of an aerial mast, the court said it saw no reason why if contractors contract in the course of their business to design, supply and erect a television aerial mast, they should not be under an obligation to ensure that it is reasonably fit for the purpose for which they knew it was intended. The question to be asked is whether the person for whom the mast was designed relied on the skill of the supplier to supply and design a mast fit for the purpose for which it was known to be required, see *Independent Broadcasting Authority* v. *EMI Electronics Ltd and BICC Construction Ltd* (1980).

### 4.3.2 Materials

Until 1995 (when the relevant statute was extended to Scotland) it was the case that there were implied by common law into contracts that required the supply of materials terms that the materials used would be of good quality and would be reasonably fit for the purpose for which they were used; that is unless it could be demonstrated that the parties' intention was to exclude the implied terms, or either of them, see *Young & Marten Ltd* v. *McManus Childs Ltd* (1968). Where the purchaser made known to the contractor the particular purpose for which the materials were required, so as to show that reliance was being placed on the contractor's skill and judgment, and where the materials were of a type which it was in the course of the contractor's business to supply, it was implied that the materials would be reasonably fit for that purpose, see *Young & Marten Ltd.*

There was a further obligation in relation to materials, namely, that they had to be free from defects which would have meant that they were not of the requisite quality. This also applies to latent defects which could have been detected using due skill and care. This did not, however, go so far as to imply a warranty by the contractor that the materials were suitable for the contract purpose in circumstances where the selection of the materials for their suitability was not a matter for the discretion of the contractor, see *Greater Glasgow Health Board* v. *Keppie Henderson & Partners* (1989).

The matter of implied terms in relation to the supply of materials is now dealt with by the Supply of Goods and Services Act 1982 which was extended to Scotland in relation to contracts made on or after 3 January 1995. Implied terms as to quality and fitness are now to be found in s.11(D)(2) and (6).

### 4.3.3 JCT 98 provisions

The contractor is obliged to use materials and workmanship of the quality and standards specified in the contract documents, by virtue of clause 2.1. If there is any discrepancy in, or divergence between, contract documents, clause 2.3 provides that the contractor is required to give immediate notice to the architect of this and the architect is then required, by clause 2.3, to issue instructions to the contractor as to how the discrepancy or divergence is to be dealt with.

By virtue of clause 8.1.1 all materials and goods are to be, so far as procurable, of the kinds and standards described in the contract bills. All workmanship is to be of the standard described in the contract bills, in the contractor's statement (if it is performance specified work) or, if no such standards are specified, of a standard appropriate to the works, see clause 8.1.2. Both of these terms are subject to the proviso that, to the extent required by clause 2.1, the materials, goods and workmanship are to be to the reasonable satisfaction of the architect.

Under clause 8.1.3, all work is to be carried out in a proper and workmanlike manner and in accordance with the health and safety plan. If the work is not carried out in this manner, clause 8.5 provides that the architect may issue whatever instructions are necessary as a result. If this is done under clause 8.5 the contractor receives no addition to the contract sum, no extension of time and no loss and expense for complying with such an instruction.

The contractor has a further obligation in relation to the quality of the works under clause 6.1. This is to comply with and give all notices required by any statutory requirement which has any jurisdiction with regard to the works. If contractors find any divergence between the statutory requirements and the contract documents they are immediately to give the architect written notice of this.

## 4.4 *Defective work*

### 4.4.1 Common law

The contractor, in addition to the obligations already considered, also remains liable for latent defects for the duration of the prescriptive period. This is on the basis that the latent defect is due to an act, neglect or default of the contractor. The contractor will have no liability if it is not possible to show a link between any breach of duty by the contractor and the loss, injury or damage incurred, if the breach and the damage caused are too remote from each other or if a term of the contract excludes the contractor's liability. The subject of prescription is considered in more detail below in Section 8.9.

### 4.4.2 JCT 98 provisions

During the currency of the contract works, the contractor can, under clause 8.4.1, be instructed to remove from the site any work, materials or goods that are not in accordance with the contract. As an alternative to this, and following consultation with the contractor and with the agreement of the employer, the architect may allow the work, materials or goods to remain. In these circumstances a deduction is made from the contract sum under the provisions of clause 8.4.2. Following practical completion, the contractor has an obligation to make good within a reasonable time the defects listed in the schedule of defects prepared by the architect in accordance with clauses 17.2 and 17.3. This schedule needs to be delivered to the contractor as an instruction of the architect not later than fourteen days after the expiry of the defects liability period.

## 4.5 *Progress of the works*

### 4.5.1 Common law

Unless it is specified in the contract that the contractor must complete by a specified date or within a specified time, the contractor is obliged to complete the works within a reasonable time. The reasonableness of the time taken is considered in light of the circumstances at the time of performance of the contract, see *H & E Taylor* v *P & W Maclellan* (1891).

The obligation to complete in a reasonable time may also come into play where there are contractual time limits set down but there has been an act or omission of the employer putting the employer in breach of the contract. For example, if the employer failed to give the contractor access to the site timeously the contractor would not be bound by the contractual time limit. In such circumstances the contractor's obligation is to complete within a reasonable time, see *T & R Duncanson* v. *The Scottish County Investment Co Ltd* (1915).

Further, if the employers are responsible in any way for the failure to achieve the completion date they are not able to recover any contractual liquidated damages from the contractor for the period of delay for which they were responsible, see *Peak Construction (Liverpool) Ltd* v. *McKinney Foundations Ltd* (1970) and *Percy Bilton Ltd* v *Greater London Council* (1982).

These rules apply even if the failure by the employers is not due to fault on their part. In a case where there were squatters on a site resulting in the employer being unable to give the contractor possession of the site, the court still held that the employer was in breach which meant the contractor's obligation to complete by the completion date changed to an obligation to complete in a reasonable time, see *Rapid Building Group Ltd* v. *Ealing Family Housing Association Ltd* (1984).

Similarly, if the employer orders extra work beyond that specified in the original contract which, as a consequence, increases the time required to complete the work, in the absence of any provisions in the contract that allow an extension of time, the employer is no longer able to claim any penalties as a result of the late completion, see *Dodd* v. *Churton* (1897). See also Section 5.5 in this regard.

Where the contractor's obligation is to complete within a reasonable time, it is still possible for employers to claim unliquidated damages (i.e. their actual losses which they would be required to calculate and prove) if the contractor takes longer than the time which would be considered reasonable in the circumstances.

It should be noted that in each of the cases referred to in this section there was no mechanism in the contract to extend the completion date as a result of the matters causing delay. Where there is such a mechanism, this would operate thereby extending the completion date so that the contractor's obligation would then be to complete by the new completion date rather than within a reasonable time.

### 4.5.2 JCT 98 provisions

The JCT provisions provide that on the date of possession the contractor shall begin the works, regularly and diligently proceed with them and complete them on or before the completion date. This obligation does not sit in isolation. There is a corresponding obligation on the employer to give the contractor possession of the site. This is contained within clause 23.1.1. The date for possession is specified in the contract.

In addition to specified commencement and completion dates, the contractor tends to work to a programme. The contractor is to provide copies of his master programme to the architect by virtue of clause 5.3.1.2. There are penalties if the contractor fails to meet the completion date in the form of liquidated and ascertained damages. These are considered below in Chapter 5. Under clause 25.3.4, contractors are subject to the overriding obligation

- constantly to use their best endeavours to prevent delay in the progress of the works,
- to prevent the completion of the works being delayed and
- to do all that may reasonably be required to the satisfaction of the architect to proceed with the works.

## 4.6 *Insurance and indemnity*

In the JCT 98 form of contract, another significant obligation owed by the contractor is in respect of insurance. Notwithstanding the existence of an indemnity by the contractor to the employer under clause 20, the contractor is obliged to take out and maintain insurance against personal injury or death and in respect of damage to property under clause 21. The JCT 98 contract also contains detailed, and particularly complex, insurance provisions under clause 22. The insurance obligations under JCT 98 are considered separately in Chapter 13 below.

## 4.7 *Health and safety*

### 4.7.1 Introduction

The issue of health and safety is a significant one in the construction industry. It is an industry that is inherently dangerous by virtue of the nature of the site environment and the operations carried out thereon. It is also an industry that has a poor record in relation to accidents. Whilst employers (in the construction sense) and sub-contractors have duties in respect of health and safety, the most significant responsibilities in this field will fall upon the contractor, in his role as an employer in the employer/employee sense.

### 4.7.2 Common law

At common law, employers (in the employer/employee sense rather than in the construction sense) have an obligation to provide

- a competent staff,
- adequate material,
- a proper system of and effective supervision and
- a safe place of work,

see *Wilsons and Clyde Coal Co Ltd* v. *English* (1938).

They also have a duty to instruct and to take steps to ensure that instructions are carried out, see *McWilliams* v. *Sir William Arrol & Co Ltd and Another* (1962).

### 4.7.3 Health and Safety at Work etc. Act 1974

The current starting point in relation to health and safety legislation is the Health and Safety at Work etc. Act, 1974. This creates duties which are incumbent on

- employers (again in the employer/employee sense as opposed to the construction sense) (see ss. 2, 3 and 9),
- employees (see s. 7),
- persons in control of premises (see ss. 4 and 5) and
- designers and manufacturers of articles and substances (see s. 6).

The principal duties incumbent upon employers, insofar as their own employees are concerned, are contained within s. 2 of the Act namely to:

- ensure, so far as reasonably practicable, the health and safety and welfare of their employees (s. 2(1))
- ensure the provision of maintenance and plant and systems of work that are, so far as reasonably practicable, safe and without risks to health (s. 2(2)(a))
- provide safe systems for the use, storage and transport of articles and substances, so far as reasonably practicable (s. 2(2)(b))
- provide such information, instruction, training and supervision as is necessary to ensure the health and safety at work of employees, so far as reasonably practicable (s. 2(2)(c))
- maintain a safe place of work and provide safe access to and egress from that place of work, so far as reasonably practicable (s. 2(2)(d))
- provide and maintain a working environment that is, so far as reasonably practicable, safe, without risks to health, and adequate as regards facilities and arrangements for employees' welfare at work (s. 2(2)(e)).

The general duty laid down by s. 2(1), and the more specific duties laid down in s. 2(2)(a) to (e) set out in statutory form the common law obligations owed by employers to their employees, see *West Bromwich Building Society* v. *Townsend* (1983).

Employers also owe duties to persons other than their employees (see s. 3). Employers need to conduct their undertaking so as to ensure that persons not in their employment who may be affected by it are not exposed to risks to their health and safety. Employers are obliged to take all reasonably practicable steps to avoid risk to people not in their employment but who might be affected by risks which arise, not merely from the physical state of the premises but also from the inadequacy of the arrangements which have been made for how the work is done, see *R* v. *Associated Octel Co Ltd*, (1996). This is particularly relevant in the construction industry where use of sub-contractors and other forms of labour, not necessarily employees of the contractor, is extremely common.

### 4.7.4 Management of Health & Safety at Work Regulations 1992

The Management of Health & Safety at Work Regulations 1992 (as amended in 1994), introduced further obligations on employers. The main obligations under these regulations require employers:

- to carry out risk assessments encompassing the identification of hazards, the type and severity of injury which may result from those hazards and the likelihood of such an injury arising (reg. 3)
- to develop preventative and protective measures and keep records of them (reg. 4)
- to provide employees with health surveillance (reg. 5)
- to appoint a competent person to facilitate the implementation of appropriate procedures in relation to health and safety matters (reg. 6)
- to keep employees appraised of health and safety issues and requirements under the Regulations (reg. 8)
- where a number of employers share a common workplace to co-operate and co-ordinate with one another in respect of health and safety issues under the Regulations (reg. 9)
- where their employees are working on another employer's site, to provide all relevant information to that employer in respect of their employees (reg. 10)
- to consider an employee's capabilities before allocating tasks (reg. 11)

### 4.7.5 Construction (Design & Management) Regulations 1994

The Construction (Design & Management) Regulations 1994 introduced new roles and duties for various parties to building contracts including clients, clients' agents (if appointed), designers and contractors. As far as the contractor is concerned, the Regulations introduced the role of principal contractor. They also introduced the role of the planning supervisor. The purpose of the Regulations was to put in place a co-ordinated structure for the management of health and safety at all stages of a construction project.

The Regulations introduced the requirement for new documents, the health and safety plan and the health and safety file.

### The health and safety plan

The health and safety plan is to be prepared at pre-tender stage by the planning supervisor (see reg. 15(1) and (2)). It is to include:

- a general description of the construction work comprised in the project (reg.15(3)(a))
- details of the timescale for the project (reg.15(3)(b))
- details of the risks to the health and safety of any person carrying out the construction work so far as these are known to the planning supervisor or are foreseeable (reg.15(3)(c))
- any other information relevant for the contractor so that he can demonstrate that he is competent to carry out the work and has allocated sufficient resources (reg.15(3)(d))
- any other information relevant for the contractor to allow him to prepare his part of the plan (reg.15(3)(e))
- any other information relevant for the contractor to allow him to comply with relevant statutory provisions (reg.15(3)(f))

The obligations of the principal contractor include the development of the health and safety plan (see reg.15(4)). This involves the contractor ensuring that the plan includes:

- Arrangements for the project to ensure the health and safety of all persons carrying out the construction work and who might be affected by it being carried out. This should take account of risks involved in the construction work including any activity which might affect the health and safety of any person carrying it out (reg.15(4)(a)).
- Sufficient information about arrangements for the welfare of persons at work by virtue of the project to enable any contractor to understand how they can comply with any requirements placed on them in respect of welfare by or under the relevant statutory provisions (reg.15(4)(b)).

### The health and safety file

The principal contractor also needs to provide all necessary information for the health and safety file. This includes:

- information about any aspect of the project or structure or materials which might affect the health and safety of any person carrying out construction or cleaning work in or on the structure at any time or any person who might be affected by this (reg.14 (d)(i)).
- any other information regarding the project which will be necessary to ensure the health and safety of any person at work who is carrying out or will carry out construction or cleaning work in or on the structure at any time or any person who might be affected by this (reg.14 (d)(ii)).

**Other obligations**

The other obligations imposed on the principal contractor by the Regulations include:

- taking reasonable steps to ensure co-operation between all contractors to enable them to comply with the requirements and prohibitions imposed on them by the relevant statutory provisions relating to the work (reg.16(1)(a))
- ensuring that every contractor and employee at work in connection with the project complies with any rules contained in the health and safety plan (reg.16(1)(b))
- taking reasonable steps to restrict access to the site to authorised persons only (reg.16(1)(c))
- ensuring that any particulars required to be included in any notice notifying the project to the Health and Safety Executive are displayed on site (reg.16(1)(d))
- promptly providing the planning supervisor with any information which he would include in the health and safety file (reg.16(1)(e)).

### 4.7.6 Construction (Health, Safety & Welfare) Regulations 1996

Also of relevance are the Construction (Health, Safety & Welfare) Regulations 1996. The salient features of the Regulations require that, insofar as is reasonably practicable, every construction site must

- provide safe and suitable access and egress,
- generally be safe for the people who work on site and
- provide sufficient working space for any party likely to work on site.

The regulations also provide guidance in relation to falling objects, namely, that reasonably practicable efforts must be made to prevent falling objects. This may include supplying, for example, a working platform or barrier.

The regulations also contain provisions in relation to

- the stability of structures,
- demolition or dismantling,

- explosives,
- excavations,
- vehicles and traffic routes,
- emergency procedures,
- welfare of persons on site and
- training inspections.

The contractor on site needs to be aware of and comply with the requirements of these Regulations.

### 4.7.7 JCT 98 provisions

Under clause 6.1.1 the contractor is under an obligation to comply with all relevant statutory requirements in relation to the works. Implicitly, this will include all health and safety legislation insofar as it is relevant to the work.

Following the Construction (Design & Management) Regulations 1994 coming into force, Amendment 14 to JCT 80 made revisions to the JCT contract to incorporate changes to ensure compliance with the Regulations. There was introduced, by way of clause 6A.2, an obligation on contractors, where they are and whilst they remain the principal contractor as defined in the CDM Regulations, to comply with all duties of a principal contractor as set out in the Regulations and in particular to ensure that the health and safety plan meets with the requirements of reg. 15(4).

Where the contractor under the building contract is succeeded as principal contractor, the contractor is obliged, under clause 6A.3, to comply, at no cost to the employer, with all reasonable requirements of the new principal contractor to the extent that those requirements are to ensure compliance with the CDM Regulations. No extension of time is to be given to the contractor for complying with this obligation.

The contractor is also to provide, and ensure that any sub-contractor provides, the information to the planning supervisor which he reasonably requires to prepare the health and safety file in accordance with regs.14(d) to (f). This obligation is contained in clause 6A.4.

# Chapter 5
# Time

## 5.1 *Introduction*

For those with a direct interest in a building project, time is an important subject. The employers will be anxious to fix when and over what period of time the works will be carried out so that they can budget and plan ahead. The contractors will be anxious to plan the commencement and carrying out of the works in order to meet their contractual obligations, express or implied, in relation to the period for completion of the works or perhaps sections or phases thereof.

A contractor's tender will normally proceed on the basis that certain operations will cost him a particular amount to carry out over a certain period of time. Generally, the longer work takes, the more expensive it is to carry out. Tendering at an appropriate level to take account of time related costs, forward planning and subsequent on-site control are essential elements of a contractor's consideration of time.

In the traditional manner in which building contracts are let in Scotland, namely, where the employer engages consultants to prepare the design and other requirements, employers should have specified what they want before the tender stage or, at least, before the formation of the contract. This places a heavy onus on the professional team of architects, engineers, services specialists and quantity surveyors. Changes after the formation of the contract should be kept to a minimum because of the effect that these are likely to have on time and cost. In the absence of agreement about the effect of such changes they may give rise to claims and disputes.

## 5.2 *Commencement of the works*

It is usual for express provision to be made for the date upon which the contractor will be given access to the site for the purpose of carrying out the works. Normally the contract will require the contractor to complete the works either by a specified date, or within a specified period from the agreed date for commencement or the date when access is given to the site. It is very important that the date of commencement of the period for completion is ascertainable and that it is specified what holidays, if any, are to be ignored in computing the period. While these are matters which common sense dictates should be made clear, experience shows that this is not always done.

## 5.3 *Time of the essence*

The phrase 'time of the essence' is one that is much used but often without much understanding. The need to do something by a specified date or time does not of itself make time of the essence. However if the contract specifically makes time of the essence, the failure to complete the works by the stipulated date amounts to a material breach of contract that entitles the innocent party to rescind and claim damages. If there is no express stipulation in the contract that time is of the essence, it can be made so by serving a notice fixing a specified time for completion. This, however, must be a reasonable one.

It is unusual for time to be of the essence in building contracts. Usually the failure of the contractor to complete is to be regarded as a breach of contract that will form the basis of a claim for damages for late completion. Even where the phrase is used, the contract must be considered as a whole. In one case the contractual clause was as follows:

> 'Time shall be considered as of the essence of the contract ... and in case the contractor shall fail in the due performance ... [the contractor] shall be liable to pay the [employer], ... liquidated damages'

In the circumstances it was nevertheless held that time was not of the essence as the contract included other terms, such as an extension of time clause, which were inconsistent with time being of the essence, see *Peak Construction (Liverpool) Ltd* v. *McKinney Foundations Ltd* (1970).

## 5.4 *Progress of the works*

### 5.4.1 Common law

As discussed in Section 5.2, the contractor's obligation is usually to complete by a particular date or within a particular period. In the absence of a relevant express contractual obligation, delay in progress prior to that date or before the end of the period probably does not in itself give the employer any rights, see *Greater London Council* v. *Cleveland Bridge and Engineering Co Ltd and Another* (1986). In such a case a claim for damages will only arise if there is actual delay in completion. In extreme cases however, particularly if combined with other failings, it may amount to anticipatory breach of contract, see *Sutcliffe* v. *Chippendale & Edmondson* (1971) and *Carr* v. *J A Berriman Pty Ltd* (1953).

Some commentators suggest that there should be an implied term that the contractor will proceed with reasonable diligence and maintain reasonable progress while others state that to imply such a term would be going too far.

### 5.4.2 JCT 98 provisions

Clause 23 of JCT 98 expressly requires the contractor to proceed 'regularly and diligently' with the works. It has been held that proceeding 'regularly and diligently' indicates a sense of activity, of orderly progress and of industry and perseverance probably such as will ensure completion according to the contract, see *London Borough of Hounslow* v. *Twickenham Garden Developments Ltd* (1971) and *West Faulkner Associates* v. *London Borough of Newham* (1994).

Moreover, clause 27 of JCT 98 gives the employer the right to determine the contractors' employment if they unreasonably suspend the carrying out of the works or if they fail to proceed regularly and diligently. Accordingly, if this right is to be exercised, it is important that the architect is satisfied that there is sufficient and reliable evidence of the contractors' failure.

## 5.5 *Extension of time for completion*

### 5.5.1 General

It is a basic principle applicable in all contracts that one party cannot seek to enforce a contractual obligation of the other party where they have prevented the other from performing that obligation. As Lord Denning put it in *Trollope & Colls Ltd* v. *North West Metropolitan Regional Hospital Board* (1973):

> 'It is well settled that in building contracts ... where there is a stipulation for work to be done in a limited time, if one party by his conduct – it may be quite legitimate conduct, such as ordering extra work – renders it impossible or impracticable for the other party to do the work within the stipulated time, then the one whose conduct caused the trouble can no longer insist upon strict adherence to the time stated. He cannot claim any penalties or liquidated damages for non-completion in that time ... The time becomes at large ... The work done must be done within a reasonable time.'

Given the complex nature of most building contracts, the need to instruct variations and to take account of unforeseen matters, it is almost inevitable that the employer will fall foul of this doctrine. To accommodate such matters, building contracts usually set out a mechanism by which the original completion date can be changed and specify the circumstances in which an extension of time for completion can be obtained. It is less common to find a contractual term which allows the contractor to make up time by way of some form of acceleration but such provisions do exist, see for example clause 3 of the JCT Management Contract.

It is a common but misguided view that extensions of time benefit only the contractor. Clearly, they give the contractor more time to complete the works and reduce or extinguish the liability for liquidated damages. However, were it not for extension of time provisions employers would not be entitled to claim liquidated damages where they have been the cause of some delay.

But dealing with an employer's default is not usually the sole reason for having extension of time clauses. Most include an entitlement to an extension of time for what might be described as neutral events. These arise through the fault of neither party, for example war, riots and bad weather. In this sense, such clauses do benefit contractors by giving them an extension of time for some matters that might otherwise be at their risk.

Indeed the terms of a particular contract may allow an extension in circumstances which some may think go beyond what might be regarded as neutral. Clause 25 of JCT 98 allows contractors to claim an extension as a result of their inability to obtain labour or materials for reasons beyond their control. While employers might seek to delete that particular provision, they should be wary of doing so. They may just provoke tenderers into increasing their tender prices to reflect the different nature of the risk.

### 5.5.2 Extensions of time under JCT 98

Clause 25 of JCT 98 contains very complex provisions detailing the circumstances in which the contractor is entitled to an extension of time. The entitlement to an extension only arises as a result of delay due to specified 'relevant events', set out below.

#### Force majeure

The term *force majeure* is thought to have been taken from the Code Napoleon. In a contract governed by Scots Law it does not have any particular technical meaning. *Force majeure* under JCT 98 is considered below at Section 8.3.2.

#### Exceptionally adverse weather conditions

Bad weather is not, in itself, a good reason for not completing on time. That means that, save where the contract contains provisions which recognise the need for an extension of time due to bad weather, the contractor will be held to have accepted the risk of completing on time notwithstanding bad weather as they are the party best able to deal with it. It is open to the parties to agree where the risk falls.

JCT 98 has taken a particular route, but the words used require careful consideration. There requires to be not just exceptionally adverse weather conditions but also delay to the progress of the works as a result.

Proof that the weather has been exceptionally adverse is usually provided by examining local weather records and comparing the actual weather experienced at a particular time of year against that of previous years at that time.

Under JCT 63 it was held to be important to note that the test is whether the weather itself was 'exceptionally inclement' so as to give rise to delay, and not whether the amount of time lost by the inclement weather was exceptional, see *Walter Lawrence and Son Ltd* v. *Commercial Union Properties (UK) Ltd* (1984). Further any delay due to weather is to be determined at the time the work is carried out, not when it was programmed to be carried out.

### Loss or damage occasioned by one or more of the specified perils

The specified perils are those listed in clause 1 of JCT 98, for example fire, lightning, explosion, storm, tempest and flood, but excluding the excepted risks which are also listed in that clause.

### Civil commotion, strikes, local combination of workmen etc.

Where there is no express provision for such matters in a contract they may be covered by general provisions relating to *force majeure* or special circumstances. A civil commotion appears to be some form of insurrection of the people that is different from a riot or a civil war. A 'local combination of workmen' is not defined in JCT 98, however, it is thought that it might cover, for example, a go-slow. JCT 98 allows the architect to take into account the possible far ranging effects of strike action. In other building contracts it may be difficult to know if an extension is to be granted only where the strike relates to onsite work or whether it extends to strikes which have an impact upon the performance of sub-contractors and suppliers. This should be made clear in the contract. Difficulties can sometimes arise, for example, in determining whether the extension should be for strikes or delay due to work to be carried out by statutory undertakers, see *Boskalis Westminster Construction Ltd* v. *Liverpool City Council* (1983).

### Compliance with certain specified instructions

There are two categories of architect's instruction that can entitle the contractor to an extension of time. Firstly, there are the matters listed in clause 25.4.5.1, which cover a wide range of circumstances where the architect is obliged to issue instructions under the contract. Secondly, an instruction requiring the opening up of the works for inspection or testing may give rise to such an entitlement, unless the inspection or test shows that work, materials or goods were not in accordance with the contract.

### Delay in receipt of necessary instructions, etc. from the architect

Delay on the part of the architect in supplying instructions, drawings, details or levels can trigger an extension of time but only provided that these were requested by the contractor on a date which, having regard to the completion date, was neither unreasonably distant from nor unreasonably close to the date on which it was necessary for the contractor to receive same.

It is very common for contractors to try to make much of the late issue of instructions, drawings and other information. In a lot of cases the issuing of drawings and information as the works progress will not cause delay provided that contractors can plan ahead and meet the contractual obligations they have undertaken. In the absence of express terms, parties would need to rely on implied terms about such matters with each asserting terms which most suited their own position. This should be avoided. JCT 98 has an express term which, although cumbersome, tries to address the situation which is truly productive of delay.

It has been held that a contractor's programme prepared at the start of the works and which set out the dates by which the contractor required to receive information and drawings may be enough, see *London Borough of Merton* v. *Stanley Hugh Leach Ltd* (1985).

**Delay on the part of nominated sub-contractors or nominated suppliers which the contractor has taken all practicable steps to avoid or reduce**

The contractor will normally be responsible for the delay occasioned by the act or omission of a domestic sub-contractor unless they are relieved of liability by an express term in the main contract (see *Scott Lithgow Ltd* v. *Secretary of State for Defence* (1989)) or a mechanism for extension of time in the main contract. Nominated sub-contractors and nominated suppliers are a different matter because the nominee and the timing of the nomination are within the employer's control. Nomination must be done timeously. Where re-nomination is required, the employer must avoid delay in so doing and must allow reasonable time for rectification of any faulty work left by the previous nominated sub-contractor.

**Delay on the part of the employer or persons engaged on their behalf to do other work associated with the works**

A building contract may provide that the employer or some other contractor on the employer's behalf will carry out work directly or indirectly associated with the works which are to be carried out by the contractor under that building contract. If it does so, there is an implied term that such work will be carried out timeously so as to allow the contractor to fulfil their obligations under the building contract. JCT 98 avoids the need to rely on such an implied term by making express provision.

**Delay caused by late supply of materials and goods which the employer has undertaken to supply**

There may be circumstances in which the employer undertakes to provide certain materials or goods for the purposes of the contract works. Should they not do so timeously the progress of the contract works may be delayed. Accordingly, these circumstances may entitle the contractor to an extension of time.

### Specified government intervention

The contractor may be entitled to an extension of time if, by reason of the government exercising a statutory power, the availability of labour, or goods and materials, or fuel and energy essential to the proper carrying out of the works is affected.

### The contractor's inability to secure labour and materials for reasons which are beyond their control and which could not reasonably have been foreseen

In the absence of provision to the contrary, the risk of being unable to secure men and materials would fall on the contractor. Some contractors seek to qualify their tenders in respect of such matters but then fail to make sure that such a qualification becomes part of the actual contract. JCT 98 contains an express provision that deals with this issue. The test of unforeseeability needs to be applied as at the time of tender. Whether that was so and whether the inability is truly beyond the contractor's control, leave scope for much argument.

### Work by local authorities and statutory undertakers

This includes such matters as electricity, gas, water and other services which need to be installed in most, if not all, buildings.

This is to be contrasted with a quite separate situation where delay to the works is caused by the need to work close to, or in physical contact with, pipes and other apparatus of local authorities or statutory undertakers. These matters will be governed by the terms of the particular contract in question. If there are no express provisions that deal with a contractor's rights in such a situation, implied terms will need to be relied upon, see *Henry Boot Construction Ltd* v. *Central Lancashire New Town Development Corporation* (1980).

### Failure by the employer to give access to the site in accordance with the contract

The contract can specify whether all of the site or parts of it will, from time to time, be made available to the contractor. This should be made clear as disputes can arise, see *Whittal Builders Co Ltd* v. *Chester-le-Street DC* (1987).

Employers should seek to avoid lengthy and costly arguments from contractors based on implied terms that will seek to make the employers responsible for all matters, including those that are strictly outwith their direct control. The authorities in this area appear to indicate that there is no implied warranty that access will not be impeded by a third party over whom the employer has no control, see *Porter* v. *Tottenham UDC* (1915), *LRE Engineering Services Ltd* v. *Otto Simon Carves Ltd* (1981) and *Ductform Ventilation (Fife) Ltd* v. *Andrews-Weatherfoil Ltd* (1995).

JCT 98 now makes specific provision where difficulties had emerged in the past, see *Rapid Building Group Ltd* v. *Ealing Family Housing Association Ltd* (1984). SMM7 requires that any conditions as to access are to be stated in the bills.

### Deferment by the employer of possession of the site

The employer must give the contractor the possession of the site anticipated in the contract to allow him to carry out the works. This is of critical importance. If he fails to do so the employer is in breach of the contract. To address such circumstances JCT 98 makes specific provision for the deferment of possession.

### Approximate quantities stated in the bills which are not a reasonably accurate forecast of the work required

This relevant event is provided to address a serious understatement in the bill of quantities of the work required. In such circumstances the contractor can be confronted with a requirement to execute far more work than envisaged at the time of entering in to the contract.

### A change in statutory requirements which requires modification of performance specified work

Statutory requirements are defined in clause 6.1.1 as any Act of Parliament, instrument, any rule or order under any Act of Parliament, or any regulation or byelaw of any local authority or of any statutory undertaker which has any jurisdiction with regard to the contract works or with whose systems the contract works are or will be connected. If any statutory requirement necessitates an alteration or modification to any performance specified work, provided the contractor has taken all practicable steps to avoid or reduce any delay occasioned by a change in the statutory requirements, he may be entitled to an extension of time.

### The use or threat of terrorism or the response of the relevant authorities to such activity

The potentially wide reaching consequences of terrorist activity, whether actual or threatened, are obvious. In recognition of the particular problems that could be occasioned by this, a new relevant event dealing with terrorism was introduced by JCT in July 1993, by way of Amendment 12 to JCT 80.

#### Compliance or non compliance by the Employer with the Clause 6A.1 requirements in relation to the CDM Regulations

This relevant event is only applicable where the CDM regulations are applicable to the contract. Clause 6A.1 requires the employer to ensure that the planning supervisor carries out all the duties incumbent upon him under the CDM regulations. If the contractor is not the principal contractor for the purposes of the CDM regulations, the employer must ensure that the principal contractors carry out all the duties incumbent upon them under the regulations.

#### Delay arising from a suspension by the contractor pursuant to clause 30.1.4

This relevant event followed upon the amendments to JCT 98 to ensure compliance with the 1996 Act. Assuming the contractor meets with the requirements of the contract in respect of suspension for non-payment (which requirements are considered below at Section 9.9.2) and thereafter suspends performance, such a suspension will be a relevant event which may entitle the contractor to an extension of time.

### 5.5.3 Contractual requirements for the granting of an extension of time

#### The contractor's notice

The building contract should set out the procedure which is to be followed when dealing with extensions of time and it is important that those involved in that process adhere to the requirements of the contract. For example, clause 25 of JCT 98 provides that whenever it becomes reasonably apparent that progress of the works is being or is likely to be delayed, the contractor should forthwith give written notice to the architect setting out the material circumstances including the cause of delay and stating which of the events is a relevant event.

It is important to note that the requirement is to give notice irrespective of whether the contractor is seeking an extension and irrespective of whether an event is a relevant event, provided it becomes reasonably apparent that progress of the works is being or is likely to be delayed. While contractors will always be reluctant to advise the architect of matters for which they are responsible, for example defective planning, poor supervision or inefficient working, the logic of this appears to be that, as the architect is only obliged to grant such extension as is fair and reasonable, it is important that he is aware of all the facts which are relevant in determining what is fair and reasonable.

It is only in respect of relevant events that the contract requires the contractor to give, in the notice or as soon as possible thereafter, particulars of the expected effects and an estimate of the extent of the expected delay.

Provisions like this often give rise to arguments about whether proper and timeous notice by the contractor is a condition precedent to an award of an extension of time. Architects and employers often argue that that is the case, but the contractual provisions in each case require careful consideration, see *London Borough of Merton v. Stanley Hugh Leach Ltd* (1985).

Contractors should always consider the terms of an extension of time clause very carefully. Should they give notice of those matters or events that they consider at the time are likely to be non-critical? Should they refrain from giving notice where they believe that they have an adequate float in terms of time to allow them to complete within the required period? Much will depend on the particular wording of the extension of time clause, but it is important to remember that things can change over the course of a contract through no fault of the contractor. In most, if not all, cases it will be prudent to give notice. A failure to give written notice of delay may, in certain circumstances, constitute a breach of contract.

It has been suggested that if the architect, because of a failure on the part of the contractors to give notice, has been unable to avoid or reduce a delay to completion, the contractors should not be awarded an extension greater than that which they would have received had they given notice, see *London Borough of Merton*.

JCT 98 provides that no extension is to be granted unless the contractor has constantly used their best endeavours to prevent delay, however caused, and they have done all that may reasonably be required to the satisfaction of the architect to proceed with the work. Unfortunately there is little guidance on what is meant by 'best endeavours' and 'all that may reasonably be required' in this context. Some take the view that the contractor must, if necessary, re-programme, increase resources and work over-time. Others take the view that, strictly, they are not obliged to take steps which would result in them incurring any material additional costs.

### Fixing a new completion date

Under JCT 98 the architect must consider if the events notified are relevant events and whether, as a result, completion is likely to be delayed beyond the completion date. Within twelve weeks of receiving the notice and reasonably sufficient particulars and estimates, and provided it is reasonably practicable for him to do so having regard to the sufficiency of such notice, particulars and estimates, the architect must fix a new completion date in writing or, alternatively, notify the contractor that he considers that no extension is due.

Although clause 25 states that the architect 'shall' act appropriately within twelve weeks, it is generally regarded that, taken in the context of the other terms of the contract, the timescale is directory only and not mandatory.

In notifying the contractor of his decision the architect must state the relevant events which he has taken into account in fixing a new completion date. However, there is no express requirement that he set out the period allowed for each event if more than one is relevant. He must also state the extent to which consideration has been had to variations giving rise to omission of work.

After practical completion has been achieved, the architect is obliged to review and reach a final view on the fair and reasonable extension of time to which the contractor is entitled, and that not later than twelve weeks after practical completion. In fixing a later completion date he can review a previous decision and have regard to all relevant events whether notified to him or not. He can only fix a completion date earlier than a previously revised completion date if that is fair and reasonable taking into account omission of work since his last decision. In no circumstances can he fix a date earlier than the original contractual date for completion.

Refusal by the architect to consider an application for extension of time may, in certain circumstances, constitute a breach of contract by the employer.

### 5.5.4 Calculation of extension of time and proof of entitlement

These matters are the subject of much controversy and a detailed examination of them is beyond the scope of this book. Indeed they appear to have generated a whole industry of consultants who profess an expertise in this area, using critical path analysis, computer technology and other techniques to provide a retrospective delay analysis which is said to be as accurate as is capable of being achieved.

### 5.5.5 Contractor's programmes

Under most forms of building contract the contractor's programme is not part of the contract. In the unusual event that the parties have agreed that the contractor's programme does form part of their contract and is binding upon them, it is of considerable significance. In such an event the contractor is obliged to work to that programme and, perhaps more significantly, the employer is obliged to allow the contractor to work to that programme. Conversely, a contractor is not obliged to work to a programme where that has not been made a requirement of the contract, see *Pigott Foundations Ltd* v. *Shepherd Construction Ltd* (1993).

Contractors are entitled to programme the works in order to complete in a lesser time than that allowed in the contract. That does not alter the obligation of the employer, which is not to impede the contractor in completing the works in the time allowed by the contract, see *Glenlion Construction Ltd* v. *The Guinness Trust* (1987) and *J F Finnegan Ltd* v. *Sheffield City Council* (1988).

However, most analyses of delay use the contractor's original programme as part of the process. Whether this can be used as a basis for any proper analysis depends upon whether the original programme was put together properly. Usually the analysis will also involve an 'as built' programme showing the actual start and finish dates for each activity specified on the programme. This allows the actual start and finish dates to be compared with those set out in the contractor's original programme. If the 'as built' programme also highlights the nature and timing of the matters upon which the contractor founds as the source of alleged delay, for example instructions and variations, suspensions and the like, such an 'as built' programme can give, at the very least, a useful picture of the factual background to the carrying out of the works.

### 5.5.6 Causation

The contractors must prove that a relevant event, and not their own inefficiencies or other matters for which they must accept responsibility under the contract, caused delay. Difficulty is caused by the fact that some contractors do not keep sufficiently detailed records of events and their impact upon the works. Rarely is it the case that there is one clear event that can be shown to be the only cause of delay to the works. If it can, there is usually little scope for real dispute. More usual is the situation where there are different events that are productive of delay some of which are the responsibility of the contractor and others which are the responsibility of the employer. They may all have an impact upon the works at the same time or they may have an impact upon the works at different times. This makes the calculation of extensions of time a very difficult area.

In such circumstances the strict application of certain legal rules relating to causation may not be possible or give rise to very unsatisfactory results. Certain rules do, however, exist. These include:

- the 'dominant cause approach', see *Leyland Shipping Co Ltd* v. *Norwich Union Fire Insurance Society Ltd* (1918) and *H Fairweather & Co Ltd v. London Borough of Wandsworth* (1987),
- the 'first in line approach', and
- the 'apportionment approach'.

Given the large number, complex nature and interaction of events on most building sites, it is considered that the extension of time to which the contractor is entitled will always be very much a matter of opinion.

## 5.6 *Partial possession and sectional completion*

There may be good reason why the parties to a building contract wish to make provision for partial possession or sectional completion of the works. The employer may need the building desperately. The contractor may wish to be relieved of obligations such as those regarding insurance and site security.

Partial possession refers to the situation where the employer takes possession of part of the works before completion of the whole. If the contract does not make express provision allowing the employer to take partial possession, he will normally be unable to do so without the consent of the contractor. Clause 18 of JCT 98 makes such provision and alters the contractor's obligations in respect of liquidated damages, insurance and defects liability. There is deemed practical completion of the part taken over by the employer for certain purposes.

Sectional completion refers to the situation where the works are defined in advance in separate sections and a different date is given for completion of each section. This requires the precise definition of each section, the date of possession, date or period for completion and liquidated damages for each section. The Joint Contracts Tribunal produces specific contractual documentation to reflect the parties' agreement to sectional completion. The relevant details for each section must be given in the abstract of conditions which is contained in the appendix to the contract. Problems arise where parties do not use a tried and tested standard form. In such circumstances they run the risk that unless great care is taken with the liquidated damages provisions, they may be inoperable.

## *5.7 Completion of the works*

### 5.7.1 Timescale for completion

If no timescale has been specified, the contractor is obliged to complete the works within a reasonable time, see *H & E Taylor* v. *P & W Maclellan* (1891). The implication of a contractual term requiring that the works should be completed within a reasonable time is most common in building operations of small value where more importance is placed on the price and specification of the work than the period within which the work is to be carried out.

Where a completion date has been agreed it may become unenforceable by virtue of some later agreement, or by waiver on the part of the employer, or where the contractor has been prevented from completing on time by acts or omissions of the employer or those for whom he is responsible. In such circumstances, unless a new completion date is agreed or the contract provides a mechanism for an extension of time, time is said to be at large. Contractors like to argue that time is at large in the sense that there is then no date by which the works must be completed. However, that is an erroneous view of what is meant by time at large. In such circumstances the contractor is still obliged to complete the works within a reasonable time.

What is a reasonable time is a question of fact, to be determined in light of all the surrounding circumstances of each particular case. If time has come to be at large, the employer is unable to recover liquidated damages under the contract because there is no fixed date for completion that can be used in the calculation of the damages. However, it is still possible for the employer to claim unliquidated damages for breach of the contractor's implied obligation to complete within a reasonable time. These matters are discussed in more detail later in this chapter.

### 5.7.2 'Practical completion'

It is perhaps trite to say that the works are complete when the contractor has executed all the work that he has contracted to perform. In building contracts, however, things are not always as simple as they might be.

Completion of the works is an important event. Most building contracts will require completion to the satisfaction of the employer, the architect or some other specified third party. It is the date of completion that is used to determine whether the contractor has completed timeously. Accordingly, where the contractor is in culpable delay it marks the end of the period for which damages for late completion are payable. Given the importance of this, most building contracts require some kind of formal certification that the works are complete.

Whether the works are complete is a matter that is ripe for dispute. If the contractor is in culpable delay and liable to pay damages for late completion they will be seeking certification of completion as soon as possible. The employer will be more concerned that the works are truly complete.

Clause 17 of JCT 98 provides for certification of 'practical completion'. The term is not defined and what it means has never been properly settled. Few matters leave so much scope for dispute. The use of the word 'practical' has been argued by some to mean something less than full completion. The generally accepted view is that a certificate of practical completion can be issued 'where very minor *de minimis* work had not been carried out' but that 'where there were any patent defects ... the architect could not have given a certificate of practical completion', see *H W Nevill (Sunblest) Ltd* v. *William Press & Son Ltd* (1981).

Given the importance of the term in JCT documentation it is disappointing that there has been no real attempt made to define precisely what it means. If definition is that difficult or agreement cannot be reached perhaps a different form of words altogether should be considered. If the wording has been left for so long in the belief that it affords some degree of flexibility, is it worth the uncertainty which is thereby created?

## 5.8 Damages for late completion

The failure of the contractor to complete the works on time as required by the contract is a breach of contract. Like any other breach of contract it gives rise at common law to the possibility of a claim for damages for that breach. The damages are determined after the breach has occurred and require proof of loss by the employer. Such a claim must also meet the requirements of the general law of damages. If the loss is too remote it will not be recoverable see *Liesboch Dredger* v. *Edison Steamship* (1933), *Hadley* v. *Baxendale* (1854), and *Victoria Laundry (Windsor) Ltd* v. *Newman Industries Ltd* (1949). The wronged party must take all reasonable steps to mitigate the loss consequent on the breach. The general law of damages is considered below in Section 9.4.

It is usually possible for employers to estimate, with a fair degree of certainty, the loss that they will sustain if the contractor does not complete on time. This can be done in a number of ways, for example by estimating additional financing costs, loss of rental and the like. For an interesting discussion of this area see *Multiplex Constructions Pty Ltd* v. *Abgarus Pty Ltd* (1992). As a result of this, and the desire of contractors to fix the level of their liability to the employer for damages in the event of late completion, most building contracts are drafted in such a way that the parties fix in advance the damages which will be payable for late completion. If these damages are a genuine pre-estimate of the loss likely to be suffered by the employer they are called 'liquidated damages'. This subject is considered below in the next section.

What is a genuine pre-estimate of loss in the context of liquidated damages is an issue which has prompted much debate, not least where the project in question is said not to be commercial in nature, see *Clydebank Engineering and Shipbuilding Co Ltd* v. *Don Jose Ramos Yzquierdo y Castaneda* (1905). That will rarely, if ever, be the case in a building contract.

There is much to be said for the view that, if the contractor does not like the liquidated damages, he should negotiate them down before entering into the contract. It is submitted that the view that liquidated damages provisions, where operable, provide an exhaustive remedy to the employer for late completion is to be preferred to the view, sometimes expressed, that it is not. The contrary view gives insufficient weight to the considerable benefits of the agreed nature of such damages.

## 5.9 Liquidated damages

### 5.9.1 General

It is normal in modern building contracts to find a liquidated damages provision to the effect that the contractor will pay or allow the employer a sum for each specified period, for example per day or per week, that the works remain incomplete after the contractual date for completion.

It is less common to find a liquidated damages provision in a sub-contract. A sub-contract may contain a provision putting the sub-contractor on notice that in the event of the main contractor's failure to complete the works timeously the employer may impose liquidated damages upon the contractor. If the sub-contractor is on notice of this, the damages which they may become liable to pay to the main contractor in the event that a breach of contract on their part causes delay to the completion of the main contract may include the amount of liquidated damages paid by the main contractor to the employer as a result of the sub-contractor's breach.

Clauses that specify liquidated and ascertained damages for delay apply where the works are completed in natural course, but not to contract time. They do not apply where the original contractor does not complete the works. They are ineffective if time has become at large since there is no fixed date from which damages can be calculated, see *British Glanzstoff Manufacturing Co Ltd* v. *General Accident, Fire and Life Assurance Corporation Ltd* (1912).

A valid liquidated damages clause removes the need for proof of actual loss, which may be difficult and costly. It should be recognised that if the clause is valid and applicable, employers are entitled to the agreed liquidated damages even if they have in fact sustained no loss. In circumstances where the liquidated damages clause is inapplicable, the employer must prove the loss which has been caused by the contractor's breach of contract.

A typical example of a provision for payment of liquidated damages is to be found in JCT 98, clause 24. This provides that, subject to the issue of a certificate of non-completion under clause 24.1, the employer may require (in writing and not later than the date of the final certificate) that the contractor pay or allow the employer liquidated and ascertained damages at the rate stated in the appendix (or at such lesser rate as may be specified in writing by the employer) for the period between the completion date and the date of practical completion.

The certificate issued under clause 24.1 is a pre-requisite to the employer's right to deduct ascertained and liquidated damages under this clause. Such a certificate may not be issued before the expiry of the period for completion or after the final certificate, see *H Fairweather Ltd* v. *Asden Securities Ltd* (1979). Notwithstanding the issue of such a certificate liquidated damages may not be payable if the liquidated damages clause does not apply, is invalid or is inoperable. If there is a dispute as to whether the employer is entitled to deduct liquidated damages from the sum certified the employer may do so, at his own risk, pending resolution of the dispute.

If a new, later, completion date is fixed after a non-completion certificate has been issued under clause 24, the certificate is superseded and a new one needs to be issued if the contractor fails to meet the revised completion date. Since Amendment 9 to JCT 80 it has been made clear that no new notice of deduction requires to be given by the employer. The employer is obliged to repay any liquidated damages already recovered for the period up to the new completion date. If the contractor should fail to complete by the new completion date the employer may not deduct liquidated damages unless a new valid non-completion certificate has been issued under clause 24, see *A Bell & Son (Paddington) Ltd* v. *CBF Residential Care and Housing Association* (1989), *Jarvis Brent Ltd* v. *Rowlinson Construction Ltd* (1990) and *J F Finnegan Ltd* v. *Community Housing Association Ltd* (1995).

If the employers' losses arising from the breach for which liquidated damages have been stipulated are greater than the stipulated amount, they are not entitled to ignore the liquidated damages clause and claim such losses as they can prove. In effect the clause operates as a limitation of the contractor's liability. Doubt remains as to whether, in the event that the liquidated damages provisions of a contract become inoperable, the employer can recover more by way of unliquidated damages than the genuine pre-estimate in the liquidated damages provisions.

Occasionally the rate of 'nil' is inserted in the appendix to JCT 98 as the rate of liquidated damages. Although it has been argued that this simply means that the parties have not agreed the sum payable by way of liquidated damages and that unliquidated damages may still be payable, the proper view is probably that it is to be treated as an agreement that no damages are to be paid to the employer for delay, see *Temloc Ltd* v. *Errill Properties Ltd* (1987). It should be noted that the reasoning in *Temloc Ltd was* not followed in *Baese Pty Ltd v. R A Bracken Building Pty Ltd* (1989). Although the contract in *Temloc Ltd* was one under JCT 80 conditions, it may have a wider application in other contracts where the terms are similar. It does not assist in determining what the position is where there is a dash (–) inserted in the appendix or the rate of damages is left blank.

It has been said that liquidated damages clauses are to be construed *contra proferentem*, (i.e. against the party putting it forward) but this requires detailed consideration of issues such as whether the parties had equal bargaining power to negotiate the contract and whether the contract is in a standard form drawn by a body on which employers, contractors and subcontractors are represented. Insofar as the clause has a limiting effect on the contractor's liability for damages for late completion, it may be capable of being challenged by the employer under the Unfair Contract Terms Act 1977.

### 5.9.2 Where liquidated damages provisions are not enforceable

A liquidated damages clause is unenforceable if the amount specified is a penalty. The classic discussion of the differences between a penalty clause and a valid liquidated damages clause is contained in the speech of Lord Dunedin in *Dunlop Pneumatic Tyre Co Ltd* v. *New Garage & Motor Co Ltd* (1915) where he said that:

> '(1) Though the parties to a contract who used the words "penalty" or "liquidated damages" may *prima facie* be supposed to mean what they say, yet the expression used is not conclusive. The court must find out whether the payment stipulated is in truth a penalty or liquidated damages ... (2) The essence of a penalty is a payment of money stipulated as *in terrorem* of the offending party; the essence of liquidated damages is a genuine covenanted pre-estimate of damage ... (3) The question of whether a sum stipulated is a penalty or liquidated damages is a question of construction to be decided upon the terms and inherent circumstances of each particular contract, judged of as at the time of the making of the contract, not as at the time of the breach ... (4) To assist this task of construction various tests have been suggested which, if applicable to the case under consideration, may prove helpful, or even conclusive. Such are (a) it will

> be held to be a penalty if the sum stipulated for is extravagant and unconscionable in amount in comparison with the greatest loss that could conceivably be proved to have followed from the breach (b) it will be held to be a penalty if the breach consists only in not paying a sum of money and the money stipulated is a sum greater than the sum which ought to have been paid ... (c) There is a presumption (and no more) that it is a penalty when a simple lump sum is made payable by way of compensation, on the occurrence of one or more or all of the events some of which may occasion serious and others but trifling damage ... On the other hand (d) it is not an obstacle to the sum stipulated being a genuine pre-estimate of damage that the consequences of the breach are such as to make precise pre-estimation almost an impossibility. On the contrary that is just the situation when it is probable that pre-estimated damage was the true bargain between the parties ...'

The authority of these rules has recently been re-affirmed by the Judicial Committee of the Privy Council, see *Philips Hong Kong Ltd* v. *The Attorney General of Hong Kong* (1993). Doubt remains in Scots Law about the enforceability of a liquidated damages clause where the works could not have been completed in the time specified, see *Robertson* v. *Driver's Trustees* (1881).

An employer is not entitled to enforce a liquidated damages clause if he has agreed not to enforce the clause or he has waived his right to do so. In the absence of contractual provision to the contrary, payment of the contract price does not constitute waiver of the right to claim liquidated damages, see *Clydebank Engineering and Shipbuilding Co Ltd* v. *Don Jose Ramos Yzquierdo y Castaneda* (1905).

# Chapter 6
# Certification

## 6.1 *Introduction*

The use of certificates in building contracts is both common and, it is submitted, essential for the proper administration of the contract. Although certificates fulfil a number of wide ranging functions, their central use is to provide triggers or mechanisms that regulate the rights and obligations of the parties to a contract during its currency and on completion. In particular, certificates often play an important role in the contractual mechanisms which regulate payment, both interim and final, progress and completion of the works and rectification of defects in the works. The existence or issue of a certificate is regularly a condition precedent to one of the parties obtaining rights in terms of the building contract.

In order to ascertain whether a building contract has any requirements for the issue of certificates, it is necessary to look at the express terms of the contract. If the express terms of the contract do require certification then the terms have to be carefully considered in order to ascertain

- what certificates need to be issued,
- the contractual preconditions which must be satisfied before a certificate can be issued, and
- the rights and obligations which flow from or are extinguished by the issue of a certificate,

see *Ata Ul Haq* v. *The City Council of Nairobi* (1962). In the absence of any express terms dealing with certification then the question of certification does not arise. The requirement for certification is not implied by operation of law.

## 6.2 *Formal requirements of certificates*

As the requirement for the use of certificates has to be expressed in the terms of the building contract, similarly the requirements as to the form of a valid certificate can also be stipulated. In the commonly used standard forms of building contract, it is unusual to find the form of certificates specified in detail and even more uncommon to find style or specimen certificates provided. However, any requirements as to form which are stipulated should be strictly followed, failing which there is a danger that the certificate will be open to challenge and ultimately held to be invalid.

If the contract is silent as to the form a certificate is to take then no particular form is required. If the certifier only has to pronounce himself satisfied in respect of certain matters then the certification may be given orally. Notwithstanding this, it is clearly preferable for certificates to be in writing, if for no other reason than to avoid evidential difficulties in subsequently proving whether certification has or has not been given.

Disputes do regularly arise in the course of building contracts as to whether certification has been given and whether or not a written document amounts to a certificate in terms of the contract. In *Halliday Construction Ltd and Others* v. *Gowrie Housing Association Ltd* (1995) a dispute arose as to whether letters written by the architect to the contractor amounted to non-completion certificates in terms of the contract. The letters simply advised the contractor that the architect had notified the employer that the contract had overrun and that liquidated and ascertained damages might be deducted. The contractor argued that such letters did not constitute certificates as they lacked the necessary form, substance and intent. The court held that the letters did amount to certificates, but only after some considerable hesitation.

In contrast, a letter written by the architect and relied on by the employer to deduct liquidated and ascertained damages was held insufficient to constitute a certificate in the case of *Token Construction Co Ltd* v. *Charlton Estates Ltd* (1973). The reasoning of the court was that it was unclear and ambiguous whether the architect had intended the letter to constitute a certificate.

In order to avoid such difficulties, when purporting to issue a certificate the certifier should make it clear that certification is being given. In this connection the use of the word 'certificate' is not essential although it is submitted that its use is prudent, see *Minster Trust Ltd* v. *Traps Tractor Ltd and Others* (1954). See also *H Fairweather Ltd* v. *Asden Securities Ltd* (1979) in which the court attached weight to the fact that a letter relied on as a certificate did not contain the word 'certify'. The use of words such as 'checking', 'approving' and 'satisfies' may not in themselves be sufficient and can give rise to ambiguity.

The certificate should leave the parties in no doubt as to its intention and effect and accordingly the rights and obligations which flow from the issue of the certificate. Where possible a certificate should refer to the relevant clause of the contract under which it is being issued and, insofar as possible, follow the wording contained in the clause.

In the event that a dispute does arise as to whether a document is or is not a certificate, then the use of extrinsic evidence may be permissible. For example, the terms of a covering letter sent with a purported certificate may provide assistance in ascertaining whether a document is truly a certificate. In *H Fairweather Ltd*, the court considered a whole course of correspondence to ascertain whether certain letters were intended to constitute certificates.

Unless otherwise required by the terms of the contract, a certificate does not have to include any reasons in support of the matters decided by or the opinions expressed in the certificate, and a lack of reasons can make the challenge of the certificate by an aggrieved party more difficult. Whether this is a perceived advantage or disadvantage is a matter for the parties to decide and take account of when drafting the contract.

In addition to any requirements regarding the form of certificates, there are other matters that should be borne in mind when preparing and issuing certificates. In particular one should ensure that any express preconditions to the issue of a certificate have been complied with. Such preconditions may include

- when the certificate needs to be issued;
- by what mechanism the certificate should be issued;
- by whom the certificate should be issued; and
- to whom the certificate should be issued.

For example, clause 30.8 of JCT 80 contains a number of preconditions to the issue of a final certificate, namely

- the end of the defects liability period,
- the issue of a certificate of completion of making good defects, and
- the sending by the architect to the contractor of an ascertainment of any loss and expense and a statement of all adjustments to be made to the contract sum.

In addition, the final certificate must be issued no later than two months after whichever of the foregoing is last to occur. Clause 30.8 goes on to stipulate what the final certificate should include being

- the amounts already stated as due in interim certificates,
- the adjusted contract sum,
- an explanation as to what the amount in the certificate relates,
- the basis on which the statement in the certificate has been calculated, and
- the difference between the two sums expressed as a balance due by the employer to the contractor or vice versa.

JCT 98 also provides in clause 5.8 that where any certificate is issued by the architect, it shall be issued to the employer and a duplicate copy immediately sent to the contractor. For a further example see *G A Group Ltd* v. *Scottish Metropolitan Property plc* (1992) where a certificate of non-completion issued prior to the expiry of the period for completion was held to be invalid. See also *Crestar Ltd* v. *Michael John Carr and Joy Carr* (1987).

It is common for standard forms of building contract to stipulate that the certificate must actually be delivered to the parties to the contract. The requirements for delivery may also be expressed including the method of delivery and the address to which delivery has to be made, for example to a limited company at its registered office. In the event that the contract does not stipulate that the certificate needs to be delivered to the parties then it is probably implied in any event. See for example the comments of Edmund Davies LJ in the case of *Token Construction Co Ltd* v. *Charlton Estates Ltd* (1973).

Minor errors in complying with any of the formal requirements of a certificate may not result in the certificate being held to be invalid provided the substance and effect of the certificate is correct and provided none of the parties to the contract have been misled or prejudiced. Nevertheless, such comfort should not be relied upon and the prudent course is to ensure that the certificate complies entirely with all the contractual requirements. If the certifier issues a certificate which is invalid, it may be open to him to reissue the certificate in a form which is valid provided he is not *functus officio* (disempowered because his role is concluded) and provided he does not alter the substance of the certificate unless the contract permits him to do so, see *Kiu May Construction Co Ltd* v. *Wai Cheong Co Ltd and Another* (1983).

## 6.3 *Interim certificates*

One of the most widely used types of certificate found in building contracts is the interim certificate, sometimes known as a progress certificate. Such a certificate is issued during the course of the contract works and is commonly designed to fulfil the dual function of monitoring the progress of the works and at the same time regulating instalment or interim payments to the contractor.

At common law, unless the contract provides otherwise, a contractor has no implied right to interim or instalment payments. Accordingly, under Scots law a contractor has no entitlement to any payment until completion of the whole of the contract works. This position has been altered in respect of some contracts by the 1996 Act; see Chapter 7 below for a fuller discussion of this point. Nevertheless, most building contracts, however, allow for interim or instalment payments to the contractor during the currency of the works.

In the commonly used standard forms of building contract, interim or progress certificates are the mechanism used as a means of regulating the timing and amount of such interim or instalment payments, see for example clause 30.1 of JCT 98 which makes provision for the issue of interim certificates by the architect. The actual intervals at which the certificates are to be issued are agreed between the parties at the time of entering into the contract and these details are inserted in the abstract of conditions found in appendix II of the Scottish Building Contract. If no period is stated then, in terms of the abstract of conditions taken with clause 30.1.3, interim certificates are to be issued monthly up to the date of practical completion and thereafter as and when further amounts are ascertained provided at least one month has passed since the issue of the previous interim certificate.

When issued, an interim certificate commonly operates in one of two ways. The interim certificate either certifies the value of work carried out at the date of the certificate and triggers payment of that amount to account of the final contract sum or, alternatively, the interim certificate certifies that the works have been completed to a particular stage triggering the release of an agreed instalment payment for that stage.

JCT 98 recognises both these alternatives. It provides a mechanism in clauses 30.2 and 30.3 for ascertaining the amount to be included in an interim certificate using the former alternative but also stipulates in the opening lines of clause 30.2 that this is subject to any agreement between the parties as to stage payments.

In terms of clause 30.7 of JCT 98, an interim certificate is also the mechanism used to certify the amounts due to nominated sub-contractors as finally adjusted or ascertained in accordance with their sub-contract conditions.

As a consequence of the fact that interim certificates are issued during the currency of the contract, the valuation of the work carried out or any assessment of the quality of the work carried out at the date of the certificate is not an exact science. Accordingly, interim certificates are not normally stipulated to be conclusive in respect of either the amount to be paid to the contractor or to the extent that they provide that the works and materials are of satisfactory quality. They simply have provisional validity, see *Beaufort Developments (NI) Ltd* v. *Gilbert-Ash NI Ltd and Another* (1998). The amount certified in an interim certificate can usually be challenged during the currency of the works. The procedure for challenging interim certificates is considered in more detail below.

In any event, the effect of interim certificates may be superseded by subsequent developments. In relation to the valuation of the work carried out this can normally be revised at the time of issue of the next interim certificate. Most standard forms of building contract allow for a revaluation of all work carried out in terms of the contract at the date of issue of an interim certificate and not simply a valuation of the work carried out since the issue of the previous certificate, see for example clause 30.2.1 of JCT 98.

With regard to the quality of work and materials, the issue of an interim certificate does not normally prevent the issue of instructions or directions in relation to remedying defective or unsatisfactory work. Indeed many defects may not be apparent at the time of issue of an interim certificate. Clause 30.10 of JCT 98 specifically provides that an interim certificate is not conclusive evidence that any work, materials or goods to which it relates are in accordance with the contract. Similarly, clause 1.5 of JCT 98 stipulates that the contractor remains wholly responsible for carrying out the works in accordance with the contract notwithstanding the fact that the value of that work has been included in a certificate for payment.

In many standard forms of building contract, the issue of an interim certificate is a condition precedent to payment of interim amounts to the contractor. In such cases if contractors do not receive a certificate then they will have no right to payment under the contract. Similarly, an employer is only obliged to pay to a contractor the amount contained in an interim certificate, see *Nicol Homeworld Contracts Ltd* v. *Charles Gray Builders Ltd*

(1986) and *Costain Building & Civil Engineering Ltd* v. *Scottish Rugby Union plc* (1994). See also the English case *Lubenham Fidelities and Investments Co Ltd* v. *South Pembrokeshire DC and Another* (1986).

Some building contracts do contain provisions whereby either party to the contract can ask the certifier to alter or review the amount of an interim certificate. Similarly, provisions are regularly found whereby either party can challenge the amount of, or the failure to issue, an interim certificate, see for example clause 41B.3 of JCT 98 as introduced by the Scottish Building Contract.

Until the coming in to force of the 1996 Act, if the employer challenged an interim certificate (and provided they could aver a genuine dispute regarding the issue of the certificate) then they could attempt to avoid making payment on the certificate, see *W & J R Watson Ltd* v. *Lothian Health Board* (1986). Employers could also attempt to avoid making payment on an interim certificate if they could rely on their common law rights of retention and set-off, or if there were any contractual rights to make deductions from amounts certified. A fuller discussion of these matters is contained below in Chapter 9. The position regarding the withholding of payment is now regulated by s. 111 of the 1996 Act. This is considered below in Chapter 7. Where contractors are aggrieved by the amount of a certificate, then normally they can also challenge the certificate. If no certificate has been issued then they may be able to raise proceedings if they can contend that the certificate has been wrongly withheld.

For many years following upon the decision in *Northern Regional Health Authority* v. *Derek Crouch Construction Co Ltd* (1984) it was the position that an aggrieved party could only challenge the amount of or the lack of an interim certificate by means of arbitration proceedings and not through the courts. In that case the parties to a building contract conferred on an arbiter the power to open up, review and revise certificates. The English Court of Appeal held that this special power had been expressly conferred on the arbiter and that the courts did not have a similar power and, accordingly, could not open up, review or revise certificates. It followed that a party aggrieved by a certificate, or the absence of a certificate, could only have recourse to the courts and sue for payment in the event that there had been fraud, bad faith or wrongful interference by the employer preventing the issue of a certificate or causing any certificate issued to be invalid. A simple failure to certify or certification of a wrong amount would not provide such a remedy.

The decision in *Northern Regional Health Authority* was followed in England and also in Scotland for fourteen years, see *D & J McDougall Ltd* v. *Argyll & Bute DC* (1987) and *Stanley Miller Ltd* v. *Ladhope Developments Ltd* (1988). As a result, great care was required when drafting arbitration clauses to ensure that an arbiter was given sufficiently wide powers to alter certificates. Similarly, great care had to be taken when deleting arbitration clauses, and it became increasingly common for parties to insert provisions in building contracts providing that the courts could open up, review and revise certificates where there was no arbitration clause. Whether such provisions competently gave the courts power to review certificates was unclear.

In 1998, however, the position altered dramatically with the decision of the House of Lords in the case of *Beaufort Developments (NI) Ltd* v. *Gilbert-Ash NI Ltd and Another* (1998) which overruled *Northern Regional Health Authority*. In *Beaufort Developments (NI) Ltd* the court held that merely because an arbitration clause gave an arbiter power to open up, review and revise certificates that did not mean that the courts could not consider the matter. The wording of such an arbitration clause did not confer on an arbiter wider powers than those enjoyed by the courts whose normal powers to enforce contracts are sufficiently wide to achieve the same result. Accordingly, interim certificates do not have binding and conclusive effect before a court and a party to a building contract can sue for payment in the courts of sums not yet certified in an interim certificate.

The reasoning behind the House of Lords decision was that the parties to the contract had conferred on the arbiter the power to open up, review and revise certificates. This illustrated that interim certificates were not intended to be binding and conclusive at all and accordingly, could not be binding and conclusive before the courts, see also *Robins* v. *Goddard* (1905). The House of Lords did stress that the position would be different in respect of certificates which are expressly stipulated to be binding and conclusive. For example, in the Institute of Civil Engineers contracts, decisions of the engineer are usually 'final and binding' unless and until the dispute or difference has been referred to arbitration and an award made and published.

Unfortunately, it is not clear from the court's decision in *Beaufort Developments (NI) Ltd* how the contractors' right to raise proceedings against the employer in the absence of a certificate is to be analysed. In particular, it is unclear what affect the decision has on the authorities referred to above in which it was held that a certificate is a condition precedent to payment of the contractor, see *Nicol Homeworld Contracts Ltd* v. *Charles Gray Builders Ltd* (1986) and *Costain Building & Civil Engineering Ltd* v. *Scottish Rugby Union plc* (1994). If the right to raise proceedings is regarded as an action to enforce payment under the contract then this would appear to be in direct contradiction to the decisions in *Nicol Homeworld Contracts Ltd* and *Costain Building & Civil Engineering Ltd* and the House of Lords might be prepared to overrule these decisions.

However, a second possible analysis of the contractors' right to raise proceedings is that it is based not on enforcement of payment under the terms of the contract but on another form of remedy. Two remedies may be appropriate, namely, a claim for payment on a *quantum meruit* basis (i.e. a reasonable amount for the work actually done) and damages. In *Costain Building & Civil Engineering Ltd* it was recognised that either of these remedies could give rise to an immediate right to payment. In *Beaufort Developments (NI) Ltd* the most helpful discussion of available remedies is found in the speech of Lord Hope of Craighead who stated:

> 'On this approach ... the court will be able to exercise all its ordinary powers to decide the issues of fact and law which may be brought before it and to give effect to the rights and obligations of the parties in

the usual way. It will have all the powers which it needs to determine the extent to which, if at all, either party was in breach of the contract and to determine what sums, if any, are due to be paid by one party to the other whether by way of set off or in addition to those sums which have been certified by the architect. It will not be necessary for it to exercise the powers which the parties have conferred upon the architect in order to provide the machinery for working out that contract ... This is because the court does not need to make use of the machinery under the contract to provide the parties with the appropriate remedies. The ordinary powers of the court in regard to the examination of the facts and the awarding of sums found due to or by either party are all that is required.'

The choice between *quantum meruit* and damages is not touched on in any of the cases we have considered in this section, or elsewhere. In building contracts *quantum meruit* has been used in cases where the contractor is unable to sue for any contractual sum as a result of his own act, see *Ramsay & Son* v. *Brand* (1898). A fuller discussion of *quantum meruit* is contained in Chapter 7 below. Where, however, the contractor is unable to sue for the contract sum as a result of an act of the employer or the certifier, the most natural analysis seems to be breach of contract. If the employer or certifier disables the contractors from recovering a sum to which they are in fact entitled, it is submitted that that amounts to a breach of contract, giving rise to an action for damages, see for example *Croudace Ltd* v. *London Borough of Lambeth* (1986). In a damages action the contract is enforced, albeit by means of a secondary remedy.

Consideration also needs to be given to whether an adjudicator has the power to open up, review and revise interim certificates. It is submitted that he must have such a power as the 1996 Act stipulates that a party must have the right to refer 'any difference' to an adjudicator for his decision, see s.108(1) to (4) of the 1996 Act. Accordingly, it is arguable that this must imply a right to open up, review and revise interim certificates. If no such right is implied and if the adjudication provisions in a building contract do not allow an adjudicator to open up, review and revise certificates then it is possible that the contract will not meet the requirements of s.108 (1) to (4) and the Statutory Scheme for Construction Contracts will apply, see s.108 (5). The scheme specifically provides that an adjudicator may open up, review and revise any decision taken or any certificate given (see Part I of the Schedule to the Scheme for Construction Contracts (Scotland) Regulations 1998, para. 20(2)(a).

The adjudication scheme found in JCT 98 also specifically provides that an adjudicator has such a power, see clause 41A.6.5.2 as introduced by the Scottish Building Contract. It is submitted that if parties drafting a building contract wish to ensure that their adjudication provisions comply with the requirements of the 1996 Act, and thus avoid any risk of the statutory scheme applying, it is prudent expressly to confer such a power upon the adjudicator.

## 6.4 Final certificates

### 6.4.1 General

A second type of certificate regularly encountered in building contracts is the final certificate. For example clause 30.8 of JCT 98 contains the mechanism and preconditions for the issue of a final certificate. The issue of a final certificate usually signals the end of the contract and can deal with a number of matters, including the final amount payable in terms of the contract which often includes any amount payable for additional or extra work. It can also include, but more normally excludes, any amounts payable in respect of damages for delay or other breaches of contract. It can mean that the contract works have been completed to the satisfaction of the certifier; that additional or extra work has been completed to the satisfaction of the certifier; and that the rectification of patent defects has been carried out to the satisfaction of the certifier.

### 6.4.2 The final certificate under JCT 98

In order to ascertain the matters that are covered by the final certificate, it is necessary to consider the express terms of the contract. It is also necessary to consider the express terms of the contract to ascertain the effect of the issue of the final certificate, see *Ata Ul Haq* v. *The City Council of Nairobi* (1962).

Clause 30.9 of JCT 98 sets out the effect of the final certificate.

- It provides that the final certificate shall have effect in any proceedings under or arising out of or in connection with the contract, by adjudication, by arbitration or by legal proceedings as conclusive evidence that where and to the extent that any of the particular qualities of any materials or goods or any particular standard of an item of workmanship was described expressly in the contract drawings, the contract bills or in any specification included in any of the numbered documents, or in any instruction issued by the architect under the conditions, or in any drawings or documents issued by the architect to be for the approval of the architect, the particular quality or standard was to the reasonable satisfaction of the architect. The final certificate is not conclusive evidence that materials or goods or workmanship comply or complies with any other requirement or term of the contract.
- It is conclusive evidence that necessary effect has been given to all the terms of the contract which require that an amount is to be added to or deducted from the contract sum or an adjustment is to be made to the contract sum, save where there has been any accidental inclusion or exclusion of any work, materials, goods or figure in any computation or any arithmetical error in any computation. In such circumstances, the final certificate is conclusive evidence as to all other computations.

- It is conclusive evidence that all and only such extensions of time, if any, as are due under clause 25 of the contract have been given.
- Finally, it is conclusive evidence that the reimbursement of direct loss and/or expense, if any, to the contractor pursuant to clause 26.1 of the contract is in final settlement of all and any claims which the contractor has or may have arising out of the occurrence of any of the matters referred to in clause 26.2 of the contract whether such claims are for breach of contract, duty of care, statutory duty or otherwise.

The issue of a final certificate is unlikely to cover claims in respect of damages for breach of contract, as it is unusual to find such matters within the certifier's remit. The issue of the final certificate can, however, preclude the employer's ability to deduct liquidated damages. Clause 24.2 of JCT 98 provides that employers may require the contractor to pay liquidated damages provided they give notice in writing prior to the issue of the final certificate, see also the decision in *Robert Paterson & Sons Ltd* v. *Household Supplies Co Ltd* (1974).

A final certificate will not have conclusive effect if the certifier has exceeded his jurisdiction or the issue of the certificate is challengeable on other grounds, for example, where the certifier has not acted independently, has acted in bad faith or has acted fraudulently. These issues are more fully considered in Section 6.6 below. Clause 30.9.1 of JCT 98 specifically provides that the final certificate will not be conclusive in the event of fraud.

### 6.4.3 The final certificate as conclusive evidence

It will be noted that the clause states on a number of occasions that the final certificate has the effect of being 'conclusive evidence' on a matter. In many standard forms of building contract, it is common to find provisions that the final certificate is to some extent conclusive and binding upon the parties. Where a final certificate is stated to be conclusive and binding and has been properly issued then its effect is final in respect of the matters covered by the certificate. Accordingly, the parties to a contract cannot challenge the certificate by adjudication, arbitration or in the courts, or ask the adjudicator, arbiter or courts to review the certificate, simply on the grounds that they are aggrieved by it or disagree with its terms. The Statutory Scheme for Construction Contracts in Scotland stipulates that an adjudicator can open up, review and revise any certificate unless the contract states the certificate is final and conclusive, see Part I of the Schedule to the Scheme for Construction Contracts (Scotland) Regulations 1998, para. 20(2)(a).

The certificate is the final expression of the certifier's decision and cannot be interfered with simply on the basis that the certificate is wrong or negligently issued, see for example *Rush & Tompkins Ltd* v. *Deaner* (1987) where, due to an error on the part of the certifier, the balance due to the contractor in terms of the final certificate was mistakenly based on sums certified rather than sums certified and paid. It is submitted that the reason for this is that the certifier has been selected from a professional discipline because he possesses and can exercise the requisite skills and knowledge when issuing certificates. It is further submitted that the certifier should have an intimate knowledge of the contract as a result of his involvement, which knowledge would not be available to an independent third party such as an adjudicator, arbiter or court. Accordingly, the certifier is the person best placed to decide any issues between the parties which fall within his remit.

### 6.4.4 The English and Scottish approaches

The matters in respect of which the final certificate is conclusive and binding differ from contract to contract and it is necessary to consider every contract on its own terms. This may not be an easy exercise particularly if the contract is not clearly drafted. Furthermore, the authorities which exist in this area are often inconsistent. The English courts have tended to interpret the conclusive effect of final certificates very broadly. In *Crown Estates Commissioners* v. *John Mowlem & Co Ltd* (1994), the English Court of Appeal considered the then wording of clause 30.9.1.1 of JCT 98. The clause under consideration by the court differed from the present wording of clause 30.9.1.1 in that it provided that the final certificate was to have effect as conclusive evidence that, where the quality of materials or the standard of workmanship were to be to the reasonable satisfaction of the architect, the same were to such satisfaction. The court held that, on a true construction of this clause, all matters of standards and quality of work and materials were for the reasonable opinion of the architect and so were concluded by the issue of a final certificate. Accordingly, if a final certificate was issued and not challenged timeously then all claims for defects arising from the standard or quality of work or materials would be defeated by the conclusive effect of the final certificate. Following this reasoning claims for latent defects not apparent at the date of issue of the final certificate would also be excluded. Similarly, claims would also be excluded for defects arising from work or materials failing to meet prescribed criteria found, for example, in the bill of quantities or specification. The case of *Colbart Ltd* v. *Kumar* (1992) is a further example of the broad interpretation favoured by the English courts.

It appears that this broad interpretation did not reflect the intention of the Joint Contracts Tribunal when it drafted clause 30.9.1.1. The intention of the Joint Contracts Tribunal was that the final certificate should only be conclusive evidence that the architect was satisfied that certain requirements of the work had been complied with where both the contractor and employer had agreed to abide by the architect's decision in respect of those requirements. It was not intended to have conclusive effect where the contractor had failed to comply with prescribed requirements of the contract documents. Following the decisions in *Colbart Ltd* and *Crown Estates Commissioners*, the Joint Contracts Tribunal revised clause 30.9.1.1 to its present form to try and reflect its original intention. Accordingly, the decisions in *Colbart Ltd* and *Crown Estates Commissioners* may now be of limited application.

In Scotland the courts appear to have taken a narrower view of the conclusive effect of final certificates. Such a view reflects the original intention of JCT. In *Firholm Builders Ltd* v. *McAuley* (1982) the court considered a clause with wording similar to that considered by the English court in *Crown Estates Commissioners*. The court held, however, that the existence of the final certificate did not necessarily defeat a claim for defective workmanship or materials. The existence of the certificate simply allowed the contractor to rely on it as conclusive evidence that, where materials or workmanship were to be to the architect's reasonable satisfaction, then the final certificate demonstrated that they were to his reasonable satisfaction. It was still open to the employer to argue that the architect should not reasonably have been satisfied.

The decision of the Court of Session in *Belcher Food Products Ltd* v. *Miller & Black and Others* (1998) also appears to support a narrower interpretation of the conclusive effect of final certificates. In *Belcher Food Products Ltd* the court attempted to distinguish *Crown Estates Commissioners* by holding that even where a final certificate was conclusive evidence of the architect's reasonable satisfaction, it did not necessarily follow that it had the further effect of being conclusive evidence that the relevant standard and quality of workmanship or materials had been achieved in a question between the employer and the contractor. In other words, the final certificate would not be conclusive evidence of the standard and quality of workmanship or materials but only conclusive evidence that the architect was satisfied with the standard and quality. It followed that the court could only decide whether the quality and standard of workmanship was satisfactory once it had heard evidence. The final certificate would, however, have strong evidential value in this connection. This was not a distinction which the court had been asked to consider in *Crown Estates Commissioners*. The court in *Belcher Food Products Ltd* further attempted to distinguish *Crown Estates Commissioners* and *Colbart Ltd* on the basis that they both concerned issues which were inherently matters for the subjective opinion of the architect. In contrast, many of the issues in *Belcher Food Products Ltd* related to whether there had been actual compliance with express contractual requirements as to the quality of materials, which could be assessed objectively.

Accordingly, the Scottish courts appear to have taken a slightly different approach to that of the English courts. The authors respectfully suggest that the approach of the Scottish courts is to be preferred. The court in *Belcher Food Products Ltd*, however, clearly felt that *Crown Estates Commissioners* and *Colbart Ltd* were sufficiently authoritative that it was necessary to distinguish them rather than openly disagree with them or refuse to follow them. The result of all this is that it is difficult to draw any general principles from the cases dealing with the conclusive effect of final certificates. Each contract has to be considered on its own terms and against the background of the particular facts which have arisen. The amendment to the clause in this regard is also of significance, the decisions in each of the cases referred to on this point being in relation to the effect of the old wording.

### 6.4.5 Challenging the final certificate

A number of the standard forms of building contract contain provisions providing that final certificates may be challenged by arbitration or other proceedings following their issue. Accordingly an arbiter is usually expressly empowered to open up, review and revise certificates, see for example clause 41B.3.3 of JCT 98, as introduced by the Scottish Building Contract. Following the decision in *Beaufort Developments (NI) Ltd* v. *Gilbert-Ash NI Ltd and Another* (1998), the courts enjoy a similar power without the requirement for any express power to be given to them. Where a final certificate can be challenged by arbitration or other proceedings then it will also be challengeable by means of adjudication under the 1996 Act. Clearly such challenges are inconsistent with the concept of final certificates having, in some circumstances, binding and conclusive effect.

The means by which this inconsistency is often dealt with in the standard forms of building contract is to provide that a final certificate does not have conclusive effect immediately on being issued, rather its conclusive effect is suspended for a stipulated period of time. During this period of time an aggrieved party is given the opportunity to challenge the final certificate by raising the appropriate proceedings. At the end of the stipulated period of time, the final certificate will have conclusive effect to the extent that it has not been challenged. In respect that the final certificate is challenged then it will still have conclusive effect subject to any award of the adjudicator, arbiter or court. In this connection see, for example, clauses 30.9.2 and 30.9.3 of JCT 98 which provide that the final certificate will not have effect as conclusive evidence provided any adjudication, arbitration or other proceedings have been commenced within twenty eight days after the final certificate has been issued.

## 6.5 *Other certificates*

In addition to interim and final certificates, a number of other types of certificate are commonly found in the standard forms of building contract. These have a wide variety of different functions. The following are some examples.

### 6.5.1 Completion certificates

These certificates are commonly used to record when the works or sections of the works have been substantially or practically completed. This can have important consequences for a number of matters including

- the start of the defects liability or maintenance period,
- the release of retention monies and
- the end of the period for which the employer is entitled to deduct liquidated damages,

see for example clause 17.1 of JCT 98 which provides for the issue by the architect of a certificate of practical completion of the works which should specify the day on which practical completion of the works is deemed to have taken place.

In terms of the abstract of conditions found in appendix II to the Scottish Building Contract, the day specified in the certificate of practical completion signals the beginning of the defects liability period referred to in clause 17.2 which runs for a period of twelve months unless the parties have agreed otherwise. It is, however, specifically provided in clause 1.5 of JCT 98 that the contractor remains wholly responsible for carrying out the works in accordance with the contract whether or not a certificate of practical completion has been issued.

### 6.5.2 Non-completion certificates

In many standard forms of building contract, as a prerequisite to the deduction of liquidated damages by the employer, the certifier needs to issue a certificate of non-completion indicating that the works are not substantially or practically complete by the date for completion agreed between the parties or any extended date thereof. JCT 98 provides for the issue of such a certificate in clause 24.1. Such a certificate is a precondition to the deduction of liquidated and ascertained damages using the mechanism set out in clause 24.2, see for example *Halliday Construction Ltd and Others* v. *Gowrie Housing Association Ltd* (1995).

### 6.5.3 Partial possession certificates

Often building contracts contain provisions whereby the employer can take possession of part of the works despite the fact that the whole of the works is not yet substantially or practically complete. This often requires the issue by the certifier of a certificate or statement identifying what part or parts of the works is or are being taken into possession by the employer. This can have important consequences for liquidated damages and protection of the works, including the question of which party is responsible for insuring the works, see for example clause 18 of JCT 98 which provides for the issue by the architect of a statement of partial possession and the effects which this has on the insurance position set out in clauses 22.3.1 and 22C.1.

### 6.5.4 Certificates of making good defects

Such certificates are regularly found in the standard forms of building contract. They are normally issued at the end of the defects liability or maintenance period when all the defects have been rectified to the satisfaction of the certifier. The issue of such a certificate often triggers the release of any remaining retentions and is usually a precondition to the issue of a final certificate, see for example clause 30.4.1.3 of JCT 98 which provides that the employer need release only half the retention percentage where work has reached practical completion but where no certificate of completion of making good defects has been issued. Similarly under clause 30.8, the issue of the certificate of completion of making good defects is a precondition to the issue of the final certificate. The provision permitting the certificate of completion of making good defects to be issued is found in clause 17.4. Clause 1.5 nevertheless provides that the issue of a certificate of completion of making good defects does not relieve the contractor of his responsibility for carrying out the works in accordance with the contract.

### 6.5.5 Challenging such certificates

The effect of the foregoing types of certificates and any other certificates which may be provided for in a building contract depends on the express terms of the building contract in question. Such certificates may or may not have conclusive effect although the position under the standard forms of building contract is that they do not normally have conclusive effect and are susceptible to challenge either by way of adjudication, arbitration or court proceedings.

## *6.6 Roles and duties of certifiers*

### 6.6.1 Who is the certifier?

In the same way that the certification process requires to be expressly stipulated in a building contract, so should the identity of the certifier. In most standard forms of building contract the certifier is a person with relevant skill and knowledge selected from an appropriate professional discipline, for example, an architect, quantity surveyor or engineer. Rather than naming an individual, it is possible to nominate a firm or partnership as the certifier. It is also possible to specify the certifier by making reference to the holder of a particular post.

The certifier named in the contract must be the person who issues the relevant certificates. Although he can delegate some of the detailed work, for example, the carrying out of measurements and calculations or the inspection of work, it is the certifier himself who must issue the certificates. He cannot delegate this function. Similarly, if a firm or partnership fulfils the role of certifier, then it must be a partner in the firm who issues the certificate although, again, work of a detailed nature may be delegated to others within the partnership, see *London Borough of Hounslow* v. *Twickenham Garden Developments Ltd* (1970). In rare cases, contracts sometimes provide that the employer is also the certifier. Such provisions are infrequent in that the contractor is unlikely to agree to them. Furthermore, such provisions are not to be recommended to either party to a contract in that there is an obvious potential for abuse. An employer fulfilling the role of certifier clearly has to act honestly and fairly, failing which his actions will be open to challenge. In addition, the courts are likely to be unsympathetic towards an employer acting as certifier particularly where certification is a precondition of payments to the contractor or where decisions require a strong element of subjectivity.

Where a certifier resigns, dies or becomes incapable of fulfilling the role of certifier then normally the contract will stipulate the procedure for making a new appointment. The Scottish Building Contract, for example, provides in clause 4 that, in the event of the certifier ceasing to be employed for the purposes of the contract, then the employer has to nominate another certifier within twenty-one days. This is subject to the contractors' right of objection within seven days of the nomination provided they have sufficiently good reason for so doing, for example, conflict of interest on the part of the new certifier. Where the contract is silent then it is submitted that there is, nevertheless, an implied term that the employer has a right to appoint a new certifier, subject to reasonable objections from the contractor. If the employers fail to appoint a new certifier then this may amount to a breach of contract on their part particularly in the event that certification is a precondition to rights on the part of the contractor, for example, to payment, see *Croudace Ltd* v. *London Borough of Lambeth* (1986). Similarly, where an employer fails to appoint a new certifier then he may be prohibited from relying on the lack of certification to defend claims by the contractor.

The certifier is normally engaged by the employer and has no direct contractual link with the contractor. The certifier's rights and obligations are set out in his terms and conditions of appointment or implied by law. Usually certifiers are members of professional bodies and their conditions of appointment may be based on standard forms issued by their professional bodies. As the certifier is normally engaged by the employer then he must act on the instructions of his employer and any acts or omissions on the part of the certifier can amount to a breach of his contract with the employer giving rise to a claim for damages against him. Furthermore, the certifier is normally also acting as the agent of the employer in respect of the building contract and any acts or omissions in this connection can place the employer in breach of contract with the contractor. In this respect, however, a difficult distinction requires to be drawn between the certifier's duties as agent of the employer and his duties as certifier.

### 6.6.2 Jurisdiction of the certifier

The jurisdiction of the certifier is defined by the express terms of the building contract. If the certifier steps beyond his jurisdiction then any certificates issued will be invalid and open to challenge by either party, see for example *Hall & Tawse Construction Ltd* v. *Strathclyde Regional Council* (1990) where the certifier made deductions in a certificate which he had no authority to make in terms of the contract. This will apply even to certificates stated to have final and conclusive effect. The certifier also needs to ensure that he has complied with any formal requirements for issuing certificates particularly those which are essential preconditions and that he has complied with any time limits set out in the contract.

Once the certifier has completed his functions then his jurisdiction falls and he is *functus officio*, having discharged his function. In many standard forms of building contract this occurs when he issues the final certificate. Thereafter, he is precluded from issuing any further valid certificates, see *H Fairweather Ltd* v. *Asden Securities Ltd* (1979). The position may differ under the Institute of Civil Engineers contracts where the engineer may have a continuing involvement under the disputes procedure following the issue of the final or maintenance certificate.

The contract can also indicate a number of other events or contingencies which will bring a certifier's jurisdiction to an end. Raising arbitration proceedings, however, will not normally make the certifier *functus officio*, see *G A Group Ltd* v. *Scottish Metropolitan Property plc* (1992). In certain instances the certifier may become disqualified. This can be a difficult area because of the dual function which certifiers are regularly required to fulfil as both agent of the employer and as independent certifier.

### 6.6.3 General duties re certification

As already indicated, the certifier is normally engaged by the employer and acts as his agent in respect of the building contract. Accordingly, he must have regard for his employer's interests and, in respect of many matters, act upon the instructions of his employer. When acting as certifier, however, he is under a duty to act fairly, honestly and independently. See *London Borough of Hounslow* v. *Twickenham Garden Developments Ltd* (1970) and *Sutcliffe* v. *Thackrah and Others* (1974).

The comments of Lord Hoffman in the case of *Beaufort Developments (NI) Ltd* v. *Gilbert-Ash NI Ltd and Another* (1998) are instructive in this regard. He stated that

> '... the architect is the agent of the employer. He is a professional man but can hardly be called independent.'

A certifier must apply the terms of the contract exercising his own skill and judgment failing which his certificates will be invalid and he will be open to disqualification. He should not be unduly influenced by or act on his clients' instructions to the extent that he jeopardises his independence and impartiality, see *Nash Dredging (UK) Ltd* v. *Kestrel Marine Ltd* (1987). Similarly, the parties are under a duty to ensure that they do not interfere with the certifier exercising his certification duties, see *Perini Corporation* v. *Commonwealth of Australia* (1969) and *Nash Dredging Ltd* v. *Kestrel Marine Ltd* (1986). A certifier does not, however, have to apply the strict rules of natural justice and he has a discretion as to how to ingather information and whether to allow parties a hearing, see *The North British Railway Company* v. *William Wilson* (1911) and *London Borough of Hounslow* v. *Twickenham Garden Developments Ltd* (1970).

In light of his dual role, correspondence or discussion of matters with the employer regarding certification will not necessarily invalidate a certificate nor will the giving of advice by the certifier to the employer. The certifier can consult the employer and the employer's advisors on legal matters that arise out of or in connection with the building contract provided any response is treated solely as advice and not as a direction or instruction. Many certifiers will take independent legal advice and it is submitted that this is the most appropriate course of action as the employer's legal advisors have their own clients' interests at heart and any advice given by them may be tainted or may unduly influence the certifier.

If the certifier does lose his independence then he will become disqualified and any certificates issued by him thereafter will be invalid. Fraud or dishonesty or taking account of unduly influential matters and advice would disqualify a certifier. Similarly, fraudulent misrepresentation on the part of either party to the contract can also invalidate certificates issued in reliance upon such a misrepresentation, see *Gray and Others (The Special Trustees of the London Hospital)* v. *T P Bennett & Son and Others* (1987) and *Ayr Road Trustees* v. *W & T Adams* (1883).

Concealed interests on the part of the certifier can also lead to disqualification. It is implied that certain interests are already known about and accepted by both parties. For example, it is known that the certifier will be paid by the employer for the work carried out by him, is liable to the employer for breach of contract and that, to an extent, he has to look after the employer's interests. Interests in the welfare of either party may, however, disqualify, for example if the certifier is a shareholder in either of the parties or has a financial interest in the outcome of the certification process. This can lead to particular difficulties where, for example, the certifier has given advice to the employer in respect of the cost of the contract. If the certifier has guaranteed a price to the employer or will receive incentive payments for savings made then this could lead to disqualification. Where, however, the certifier has simply estimated the cost of the job for the employer then it is unlikely that this would amount to an interest sufficient to disqualify.

A certifier may also be disqualified if his dual roles as employer's agent and certifier become incompatible. This may occur where the certifier becomes a witness to fact or a key witness in support of one party's position.

A certifier is not disqualified simply because of an error in judgment or an error in the exercise of his discretion. Where such errors occur then it may be possible for one party to the contract to challenge the certificate by means of adjudication, arbitration or through the courts. There is authority which suggests that there is no obligation on an employer who becomes aware of errors on the part of the certifier to bring them to his attention and ensure that he adequately performs his duties, see *Lubenham Fidelities and Investments Co Ltd* v. *South Pembrokeshire DC and Another* (1986). This decision should be contrasted with the decisions in *Perini Corporation* and

*Panamena Europea Navigacion (Compania Limitada)* v. *Frederick Leyland & Co Ltd* (1947).

Failures by the certifier can place him in breach of contract with his employer and may result in the certifier being liable to the employer in damages, see for example *Jameson* v. *Simon* (1899) where the certifier was found liable in damages to his employer for certifying work which disconformed to the contract as a result of failing to exercise reasonable supervision. See also *Sutcliffe* v. *Thackrah and Others* (1974).

Such failures can also place the employer in breach of contract with the contractor, although the certifier himself would have no liability to the contractor as there is no direct contractual link between certifier and contractor. The certifier could only be liable to the contractor in circumstances where he has procured the employer to breach his contract with the contractor, see *John Mowlem & Co plc* v. *Eagle Star Insurance Co Ltd and Others* (1992), or in delict. A simple under-certification may not be sufficient to give rise to delictual liability, see for example *Pacific Associates and Others* v. *Baxter* (1988), *Lubenham Fidelities and Investments Co Ltd,* and *Leon Engineering & Construction Co Ltd* v. *Ka Duk Investment Co Ltd* (1989). It should be remembered that the certifier is not independent in the same way that an adjudicator, arbiter or court needs to be independent as he does have a clear dual function to perform. When fulfilling the role as certifier, however, he must act fairly, honestly and in good faith and must be capable of forming his own independent view on the issue requiring certification.

# Chapter 7
# Payment

## 7.1 *Contractual payment*

### 7.1.1 Introduction

One of the main obligations owed by the employer under a building contract is to make payment to the contractors for the work carried out by them. The building contract will ordinarily contain express provisions relating to payment and the parties are, subject to the provisions of the 1996 Act, free to agree between them

- the sum that is to be paid for the works,
- whether instalment payments are to be made,
- when payments are to be made, and
- the mechanism or procedure to facilitate payment.

The commonly used standard forms of building contract contain detailed provisions in respect of payment, see for example clause 30 of JCT 98.

### 7.1.2 When are payments due?

#### Traditional Scots law

In any building contract one of the contractor's prime concerns is the timing of payments for work carried out. Traditionally under Scots law contractors have no implied right to payment until they have completed all the work they have contracted to carry out, see *Muldoon* v. *Pringle* (1882) and *Readdie* v. *Mailler* (1841). It is a general principle of Scots law that, in the absence of any provisions in a contract providing for interim or instalment payments, there is no obligation to make payment until the entire contract has been fulfilled. Whether or not this applies to building contracts was considered in the case of *Charles Gray & Son Ltd* v. *Stern* (1953). In this case, the contractor demanded payments to account while building a house, maintaining that it was normal building trade practice to pay contractors for nine-tenths of the work completed. The contract did not expressly provide for interim payments to be made. The court held that the contract was a lump sum contract and that, as the contractors had failed to carry out a material part of the works, they could not sue for payment under the contract.

**Exceptions to the basic principle**

However, whilst it may be the general principle that contractors are not entitled to payment from an employer before they have carried out all their obligations under the contract, there are situations in which the contractors may be entitled to payment notwithstanding the fact that they have failed to complete.

In certain instances, if the work can be said to be 'substantially complete', the contractor may be entitled to payment of the whole contract sum, less payment for that part not completed or, alternatively, the cost of remedying any defects in the works. In *Ramsay & Son* v. *Brand* (1898) Lord President Robertson stated that:

> 'A building contract by specification necessarily includes minute particulars, and the law is not so pedantic as to deny action for the contract price on account of any and every omission or deviation. It gives effect to the principle [that if builders choose to depart from the contract they lose their right to sue for the contract price] by deducting from the contract price whatever sum is required to complete the work in exact compliance with the contract.'

Similar authority can be found in the cases of *Speirs Ltd* v. *Petersen* (1924), *Hoenig* v. *Isaacs* (1952) and *Stewart Roofing Co Ltd* v. *Shanlin* (1958).

Following the case of *Forrest* v. *The Scottish County Investment Company Ltd* (1915), it appears that the contractor's right to payment where there are defects may be stronger in a measurement contract than a lump sum contract, see Section 7.1.6 below for a definition of these different types of contract. This approach also receives some support in *Speirs Ltd.*

If, however, the contractor's deviation from the contract is material then he will be prevented from claiming payment under the contract, see *D. Ramsay & Son* and also *Dakin & Co Ltd* v. *Lee* (1916). Unfortunately, it is not always clear whether a deviation is material and this will require to be considered in light of the facts of each individual case, see for example *McMorran* v. *Morrison & Co* (1906) This can include consideration of the dimension, complexity and value of the contract and consideration of the cost of rectifying the deviation relative to the whole contract price, see *Speirs Ltd.* On the other hand, if the employer has benefited from the work carried out by the contractor then the contractor may still be able to claim compensation under the principle of *quantum lucratus* even where there are material deviations. A fuller discussion of this principle is contained in Section 7.5 below.

Where the contractors are prevented from completing the contract works by a matter outwith their control then they may still be able to claim payment for the work carried out. For example, where the employer denies the contractor access to the site to complete the contract works then the contractor will be able to claim payment for the work he has carried out. Similarly, the contractor will be entitled to claim payment for all work executed in accordance with a contract in the situation where the works are destroyed and cannot be completed, see for example the cases of *Andrew McIntyre and Company* v. *David Clow and Company* (1875) and *Richardson* v. *County Road Trustees of Dumfriesshire* (1890). This may, however, depend upon which party has assumed the risk of damage to the works during their construction and careful consideration may need to be given to any insurance provisions found in the contract. A fuller discussion of insurance is contained in Chapter 13 below.

Finally, and most importantly, the contractor may be entitled to claim payment prior to completing the contract works where the express terms of the building contract specifically provide for the making of instalment or interim payments. In modern construction contracts such provisions are almost always included, as most contractors would be unable to fund internally the ongoing construction costs pending payment upon completion. Procuring external funding would significantly increase both costs and complexity.

### 7.1.3 Payment under the 1996 Act

In respect of building contracts entered into after 1 May 1998, the position in relation to payment has altered dramatically following the coming into force of the 1996 Act. This Act materially improves the rights of the party to be paid under a construction contract as compared to the common law position.

#### Sections 109 and 110

Section 109 of the Act deals with payment and provides that a party to a 'relevant construction contract' is entitled to payment by instalments, stage payments or other periodic payments in respect of work carried out under the contract. Where the parties are entitled to interim payments under s.109 then the 1996 Act provides that the parties are free to agree the amounts of such payments and the intervals or circumstances in which they will become due.

Section 110(1) of the 1996 Act stipulates that every construction contract must provide a mechanism for determining what payments become due and when they become due and provide a final date for payment in relation to any sum which becomes due. The parties can, however, agree the interval between the date on which a sum becomes due and the final date for payment.

Section 110(2) stipulates that every construction contract must provide for the giving of notice by a party not later than five days after the date upon which a payment becomes due under the contract, or would have become due if the other party had carried out its obligations under the contract and no set-off or abatement was permitted by reference to any sum claimed to be due under one or more other contracts, specifying the amount (if any) of the payment made or proposed to be made, and the basis upon which that amount was calculated.

### The Scheme for Construction Contracts (Scotland) Regulations

Should the parties fail to provide for all or any of the matters which are required in terms of s.109 and/or s.110 then the payment provisions found in Part II of the Schedule to the Scheme for Construction Contracts (Scotland) Regulations 1998 will apply to the extent required to cover any matter not otherwise agreed.

The Schedule deals firstly with 'relevant construction contracts' being any construction contract other than one which specifies that the duration of the work is to be less than 45 days or in respect of which the parties agree that the duration of the work is estimated to be less than 45 days. For the definition of a 'construction contract' for the purposes of the 1996 Act, see Section 1.2.2.

Paragraph 4 provides that interim payments under a relevant construction contract will become due when the payee has made a claim and seven days have expired following the relevant period. Paragraph 12 provides that the 'relevant period' is as specified in or calculated by reference to the contract, failing which it is a period of 28 days.

Paragraph 5 provides that the final payment under a relevant construction contract, namely the difference between the contract price and the aggregate of any interim payments, will be due when the payee has made a claim and 30 days have expired following completion of the work.

In respect of contracts where the duration of the work is less than 45 days or where the parties agree that the duration of the work is estimated to be less than 45 days then paragraph 6 of the Schedule provides that payment of the contract price shall become due when the payee has made a claim and 30 days have expired following completion of the work. Paragraph 7 provides that any other payment under a construction contract will be due when the payee has made a claim and seven days have expired following the completion of the work to which the payment relates.

Paragraph 8 provides that the final date for any payment under a construction contract will be 17 days from the date when the payment becomes due. Finally, paragraph 9 provides a mechanism for the giving of notice by the paying party in accordance with the requirements of s.110(2) of the 1996 Act, which requires the paying party to specify how a payment is calculated.

## 7.1.4 Payments under JCT 98

The payment provisions contained in clause 30 of JCT 98 have been revised to ensure that they comply with the requirements of the 1996 Act. Clause 30.1 provides for the issuing of interim certificates by the architect and the final date for payment is stated to be 14 days from the date of issue of each certificate. Failure by the employer to make payment by this date now entitles the contractor to simple interest on the amount not paid, to be calculated at the rate of 5% over the Bank of England Base Rate. This is considered below at Section 9.6.3.

The contract leaves it to the parties to agree the intervals at which certificates will be issued and the agreed interval should be inserted in the Abstract of Conditions found in Appendix II of the Scottish Building Contract. In the absence of such agreement, the Abstract of Conditions and clause 30.1.3 provide that interim certificates are to be issued monthly up to the date of practical completion and thereafter as and when further amounts are ascertained, provided at least one month has passed since the issue of the last interim certificate.

Clause 30.12 of JCT 98, as amended by the Scottish Building Contract, provides that should the architect not issue any interim certificate then the contractors are still entitled to make an application to the employer for payment of any sums they consider are due under the contract. This right of application can only be used prior to practical completion and is to be made in accordance with any valuation statement prepared by the quantity surveyor under clause 30.1.2.2 or, in the absence of such a statement, in accordance with any application submitted by the contractors to the quantity surveyor setting out what they consider to be the amount of their valuation.

For the purposes of clause 30.12, the final date for payment is stipulated to be 14 days from the date of issue of the application by the contractor to the employer. The employer has five days from receipt of the application to give written notice to the contractor specifying the amount of the payment proposed to be made, the basis on which the amount is calculated and to what the amount relates.

This mechanism under clause 30.12 is designed to provide contractors with a means of obtaining payment in a situation where they do not receive an interim certificate. Accordingly, JCT 98 as amended by the Scottish Building Contract, the issue of an interim certificate is no longer a condition precedent to payment of any interim amounts to the contractor. See Section 6.3 for a fuller discussion of this point.

The timing of the final payment to be made to the contractor under JCT 98 is also governed by the provisions of clause 30. Clause 30.6 stipulates that, not later than six months after practical completion, the contractor has to provide all documents necessary for the purposes of adjustment of the contract sum. The architect, or, if he instructs, the quantity surveyor, then has three months to prepare a statement of all adjustments to be made to the contract sum and to ascertain any loss and expense due to the contractor. This statement and ascertainment has to be sent to the contractor forthwith upon preparation. In terms of clause 30.8, once the statement and ascertainment have been sent to the contractor under clause 30.6, the defects liability period has ended and a certificate of completion of making good defects has been issued, then the architect shall within two months thereafter issue a final certificate. The final certificate should state the final contract sum, the amounts already certified in interim certificates and the difference between the two expressed as a balance due to the contractor by the employer or *vice versa*. Clause 30.8.3 provides that the final date for payment of this balance is 28 days from the date of issue of the final certificate.

### 7.1.5 Pay when paid

Prior to the coming into force of the 1996 Act, the timing of payments under building contracts was often stipulated to be dependent upon the receipt of funds by the paying party from another source. The most common example was for main contractors to make it a condition of a sub-contractor's entitlement to payment that the main contractor had in turn received payment from the employer under the main contract. See *Taymech Ltd* v. *Trafalgar House Construction (Regions) Ltd* (1995). Such provisions, commonly known as 'pay when paid' clauses, have now been outlawed by the 1996 Act, except in very limited circumstances. Section 113 provides that any provision which makes payment under a construction contract conditional on the payer receiving payment from a third person is ineffective unless that third person is insolvent. Where a payment provision is ineffective then the relevant provisions found in Part II of the Schedule to the Scheme for Construction Contracts (Scotland) Regulations will apply. It is submitted, however, that it may still be open to parties to a building sub-contract to agree to a provision that makes payments to a sub-contractor conditional upon the value of work carried out by the sub-contractor being valued and included in a certificate issued under the main contract. See Chapter 10 for a fuller discussion of sub-contracts.

### 7.1.6 The amount to be paid

The final contract price, or a mechanism for ascertaining the final contract price, is a fundamental and essential part of the contract and should be agreed at the time the parties enter into the contract. See *Uniroyal Ltd* v. *Miller & Co Ltd* (1985). If no price or mechanism has been agreed then it may be possible for a contractor to obtain payment on the basis of *quantum meruit*. This is discussed in Section 7.4 below.

The price to be paid by the employer to the contractor for carrying out the contract works may be ascertained by a number of different methods. In many building contracts it may not be possible to calculate the contract price until after completion of the works. See for example, *Arcos Industries Pty Ltd* v. *The Electricity Commission of New South Wales* (1973).

#### Lump sum contracts

In what are commonly known as lump sum contracts the employer and the contractor agree the price for the contract works at the time of entering into the contract. Assuming the contractors complete the contract works then they will be entitled to be paid the agreed price regardless of what the works have actually cost to construct. See, for example, *Mitchell* v. *Magistrates of Dalkeith* (1930). Even in lump sum contracts, however, the price can alter as a result of a number of matters including

- additions to or omissions from the contract works,
- events giving rise to loss and expense, and
- fluctuations in cost.

An error on the part of the contractor in calculating the price will not, however, result in an alteration to the price and the contractors are bound by the price even if the work costs more than they allowed for. See *Seaton Brick and Tile Company Ltd* v. *Mitchell* (1900). A contractor will only be entitled to additional payment if he has a contractual entitlement thereto.

Under JCT 98, as amended by the Scottish Building Contract, the employer and contractor agree a contract sum at the time of entering into the contract and clause 2 of the Scottish Building Contract provides that the employer shall pay to the contractor the contract sum or such other sum as shall become payable in accordance with the conditions of contract. Clause 14.2 further provides that the contract sum shall not be adjusted otherwise than in accordance with the express provisions of the conditions and, subject to clause 2.2.2.2, any error (whether arithmetic or not) in the computation of the contract sum shall be deemed to have been accepted by the parties. Clause 2.2.2.2 provides for the correction of errors in the preparation of the contract bills (which may be errors of quantity or description) and stipulates that such a correction is to be treated as a variation. The provisions in the conditions governing final adjustment of the contract sum are found in clause 30.6. A fuller discussion of the matters which can give rise to adjustment of the contract price is contained below in Section 7.2.

### Measurement contracts

In what are commonly known as measurement contracts, no agreed price is ascertainable prior to the carrying out of the works. The price is calculated by measurement of the work actually carried out during the currency of and on completion of the contract and this work is then valued by applying a schedule of rates agreed at the time of entering into the contract. The schedule of rates will often take the form of a bill of quantities. Very basic forms of measurement contracts can be seen in the cases of *Jamieson* v. *McInnes* (1887) and *Wilkie* v. *Hamilton-Lodging House Company Ltd* (1902). Measurement contracts are often used where it is impossible to ascertain the full extent of the contract works at the time of contracting, or where there is insufficient information available to do so. For example in a contract to construct a road it may not be possible, at the time of contracting, to ascertain the exact ground conditions which the contractor will encounter.

### Reimbursement contracts

A third method often employed to ascertain the price to be paid is that used in what are commonly known as reimbursement contracts. In such contracts contractors are paid for the cost of the work carried out by them, normally with an additional allowance for overheads and profit or, alternatively, a fee for managing the contract. The additional allowance to be paid to the contractor is normally agreed at the time of entering into the contract and is usually a specified sum or, alternatively, an agreed percentage of the total contract price. It is important when drafting reimbursement contracts to ensure that contractors only recover costs properly and reasonably incurred in order to avoid extravagance or inefficient working on their part.

### Interim payments

Where the contractor has an entitlement to receive interim payments the contract will normally contain a mechanism for ascertaining the amount of such payments. This often involves the issue of interim certificates, discussed above in Section 6.3. Interim payments are payments to account of the final contract sum and normally represent either an agreed instalment payment due on a particular date, or an agreed instalment due at completion of a particular stage of the works (see, for example, *The Government of Newfoundland* v. *The Newfoundland Railway Company and Others* (1888)), or a valuation of the work carried out at a particular date (see, for example *F R Absalom Ltd* v. *Great Western (London) Garden Village Society Ltd* (1933).

Interim payments are often subject to review by later payments and on completion of the contract, see, for example, *The Tharsis Sulphur and Copper Company Ltd* v. *McElroy & Sons* (1878) in which Lord Chancellor Cairns stated that 'payments made under [interim certificates] are altogether provisional, and subject to adjustment or readjustment at the end of the contract'. See also *Beaufort Developments (NI) Ltd* v. *Gilbert-Ash NI Ltd and Another* (1998). Thus interim certificates are not usually conclusive as to either the value of work carried out at the date of payment or the quality of work carried out.

It has, however, been suggested that it is for the employer to prove that an interim valuation is inaccurate. In *Johnston* v. *Greenock Corporation* (1951) Lord Sorn stated:

> '... [T]he proper time for an employer to challenge any item in the contractor's monthly account is at the time he receives it, and when he is checking it with a view to payment. That is the time when the facts are fresh in the contractor's mind and when he can best give explanations, or make the necessary inquiries or investigation into any matter which requires explanation. At that stage the onus is clearly on the contractor to justify and explain every item in his account if called upon to do so. But if the accounts are checked, and all explanations asked for having been

satisfactorily given payment is made on them, so that the contractor naturally thinks that the business is over and done with, and then, after the lapse of a year or two, the employer seeks to reopen particular items in the account I think the situation is different ... [W]here objections can be so infinitely varied in character, it may not be advisable to attempt to lay down any general rule about onus, but it is at least clear that great care must be taken to see that the contractor is not prejudiced by the delay in bringing forward the challenge. Perhaps it would not be going too far to say that, instead of it being for the contractor to justify his charge, it is, at least initially, for the employer to show why the item, which he had already passed and paid for, should not stand.'

**Section 110(1)**

Section 110(1) of the 1996 Act stipulates that every construction contract must provide an adequate mechanism for determining what payments become due under the contract. If the contract fails to do so then the relevant provisions of Part II of the Schedule to the Scheme for Construction Contracts (Scotland) Regulations will apply. Paragraph 2(1) provides that the amount of any interim payments shall be the difference between the amount determined in accordance with paragraph 2(2) and the amount determined in accordance with paragraph 2(3).

The amount determined in accordance with paragraph 2(2) is the aggregate of

(a) an amount equal to the value of any work performed in accordance with the contract from commencement of the contract to the end of the relevant period;
(b) where the contract provides for payment for materials, an amount equal to the value of any materials manufactured on site or brought onto the site from commencement of the contract to the end of the relevant period; and
(c) any other amount which the contract specifies shall be payable from commencement of the contract to the end of the relevant period.

The relevant period is defined in paragraph 12 as the period specified in, or calculated by reference to, the construction contract or (where no period is specified or so calculable) a period of 28 days. The amount determined in accordance with paragraph 2(3) is the aggregate of any sums which have been paid or are due for payment by way of interim payments during the period from commencement of the contract to the end of the relevant period. It is further provided in paragraph 2(4) that the amount of any interim payment shall not exceed the difference between the contract price and the aggregate of the interim payments which have become due.

### Section 110(2)

Section 110(2) of the 1996 Act provides that every construction contract shall provide for the giving of notice by a party not later than five days after the date on which a payment becomes due from them or would have become due if (a) the other party had carried out their obligations under the contract, and (b) no set-off or abatement was permitted. The notice must specify the amount (if any) of the payment made or proposed to be made and the basis upon which that amount was calculated. If a contract does not contain such a provision then paragraph 9 provides a mechanism for the giving of such notice.

### The JCT 98 provisions

JCT 98 contains complex provisions for ascertaining the amount of interim payments. These provisions have been amended to ensure compliance with s.110 of the 1996 Act. The procedure for ascertaining amounts to be included in interim certificates is to be found in clause 30.2. This stipulates that the amount stated as due in an interim certificate, subject to any agreement between the parties as to stage payments, shall be the gross valuation of the work carried out by the contractor less any amount stated as due in interim certificates previously issued and less any amount which may be deducted and retained by the employer by way of retention under clause 30.4. Section 7.6 below contains a fuller discussion of retention. The provisions for ascertaining the gross valuation of the work carried out by the contractor are to be found in clauses 30.2 and 30.3.

Interim valuations are made by the quantity surveyor whenever the architect considers them necessary for the purpose of ascertaining the amount to be stated as due in an interim certificate. The contractor can also require the quantity surveyor to make an interim valuation by submitting an application under clause 30.1.2.2. Such an application must be submitted no later than seven days before the date of an interim certificate and should set out what the contractor considers to be the amount of his gross valuation. To the extent that the quantity surveyor disagrees with the gross valuation in the contractor's application then the surveyor shall at the same time as making his valuation submit to the contractor a statement which identifies the nature of that disagreement.

When the architect issues interim certificates he states

- the amount due to the contractor by the employer,
- to what the amount relates, and
- the basis upon which the amount is calculated.

Not later than five days after the date of issue of an interim certificate the employer must give a written notice to the contractor which specifies

- the amount of the payment proposed to be made,
- to what the amount of the payment relates, and
- the basis upon which that amount is calculated.

Contractors will, accordingly, know how much they are to be paid and how that sum has been arrived at.

In the absence of an interim certificate, the contractor can still make application to the employer for payment under clause 30.12.1. This application should be for an amount in accordance with the statement prepared by the quantity surveyor pursuant to clause 30.1.2.2 or, in the absence of such a statement, in accordance with the contractor's application submitted to the quantity surveyor.

## *7.2 Adjustment of the contract price*

### 7.2.1 Introduction

As indicated above, the final contract price or a mechanism for ascertaining the final contract price should be agreed at the time of entering into the contract. Even where such an agreement has been reached, matters can still arise during the carrying out of the works which will result in either an increase or a decrease in the price. The most common matters which give rise to an adjustment in the price are

- contractual variations,
- fluctuations in cost, and
- claims for direct loss and expense.

There are, however, many other matters which can potentially give rise to an adjustment in the price.

Most building contracts provide either expressly or by implication, that the contractor is obliged to carry out, and the employer is obliged to pay for, the work which the parties have agreed will be carried out. Failure by either of the parties will amount to a breach of contract giving rise to the possibility of the contract being brought to an end (as discussed below in Chapter 8) and also to other remedies becoming available to the innocent party (as discussed in Chapter 9).

If the works cost more than the contractor priced for they are still bound by the price or the mechanism for ascertaining the price which was agreed between the parties, see *Seaton Brick and Tile Company Ltd* v. *Mitchell* (1900). Similarly, the employer is so bound even if the works cost the contractor less than the price agreed or the price ascertained using the agreed mechanism, see *Mitchell* v. *Magistrates of Dalkeith* (1930). In many instances, however, the parties will attempt to claim an increase or a decrease in the contract price as a result of work they claimed was added, omitted or varied from the original agreed scope of the works.

If contractors claim additional payment for work which actually formed part of the original contract works then clearly they will have no entitlement to such additional payment. If claims are made in respect of work which was added, omitted or varied in terms of the express provisions of the contract then there may be an entitlement to an adjustment of the contract price, see section 7.2.2 below. The most difficult situation, however, is where there are no express provisions dealing with such matters.

In such cases, the removal by the employer of any work from the contractor may amount to a breach of contract on the part of the employer unless the work is paid for. If the contractor does not agree to the omission and price reduction then he may have a valid claim against the employer for his loss of profit on the portion of the work which has been omitted.

Similarly, an unauthorised variation by the contractor may amount to a breach of contract on the part of the contractor. Where the deviation is material the contractor may have difficulty claiming payment not only for the cost of the varied work but also for the work he has executed in accordance with the contract, see *Ramsay & Son* v. *Brand* (1898) and the related cases referred to in section 7.1.2. Obviously if the contractor can show that the variation was agreed to by the employer then it will not amount to a breach of contract and the contractor will be entitled to be paid for the varied work. It appears that such agreement may be established by words or conduct on the part of the employer or his agent, see *Holland Hannen & Cubitts (Northern) Ltd* v. *Welsh Health Technical Services Organisation and Others* (1981). Contrast, however, the decision in *Burrell & Son* v. *Russell & Company* (1900). In a situation where a variation has been agreed to by the employer because it assists or is convenient to the contractor then the contractor will not be entitled to any additional payment, see *The Tharsis Sulphur and Copper Company Ltd* v. *McElroy & Sons* (1878).

Where the contractor executes additional work, he will have no entitlement to payment unless he can show that the employer agreed to pay for the work. This is a general principle of Scots Law which does not apply only to building contracts. For example, in *Walter Wright & Co Ltd* v. *Cowdray* (1973) an electrical contractor was instructed to dry out and test two motors. In addition, the contractor carried out certain repairs to the motors which had not been instructed. The Court held that the contractor had no entitlement to payment for the performing of the repairs. See also *Wilson* v. *Wallace and Connell* (1859). In such situations the only means of obtaining payment may be by using the principle of *quantum lucratus* which is discussed below in section 7.5.

It is relatively common in building contracts to find provision for other payments to be made to the contractor in addition to the contract price and sometimes the parties will agree after the contract has been concluded and the works have commenced that additional payments are to be made.

In many of the standard forms of building contract provision is made for the opening up for inspection or testing of work carried out by the contractor. It is usually provided that the contractor will be paid for all such work, unless the opening up discloses that work carried out is not in accordance with the contract, see for example clause 8.3 of JCT 98. Similar provisions are to be found in ICE contracts, see for example *Hall & Tawse Construction Ltd* v. *Strathclyde Regional Council* (1990).

Many of the standard forms of building contracts, for example clause 6.2

of JCT 98, provide that the contractor is required to pay any fees or charges in respect of the contract works demandable under any Act of Parliament, byelaw or similar regulation. Under certain contracts, including JCT 98, these payments may be reimbursed to the contractor in addition to the contract price. Similarly, clause 9 of JCT 98 provides that where a contractor incurs liability by way of royalties, damages or other monies as a result of any infringement or alleged infringement of any patent right arising from compliance with an architects' instruction then any such liability will be added to the contract sum.

Where the employer wishes completion earlier than contracted for, or where the works have been delayed, then the employer may agree a bonus payment to the contractor if they achieve early or timeous completion of the works, see for example *Williams* v. *Roffey Bros & Nicholls (Contractors) Ltd* (1990). In such cases, the circumstances in which the payment is to be made should be clearly set out. In particular, it should be made clear whether the payment is truly a bonus in the sense that it is in addition to any other right to payment which the contractor may have in terms of the contract between the parties, for example direct loss and expense. It should also be made clear what is to happen to such payment if the works are delayed by subsequent variations instructed by the employer.

In addition to the contract price the contractor will also be entitled to payment of any value added tax which is applicable. Normally, the agreed contract price is exclusive of value added tax, see for example clause 15 of JCT 98. In certain circumstances the contractor may also be entitled to interest on the contract price where the employer has failed to make payments timeously in terms of the contract. Clause 30.1.1.1 of JCT 98 expressly stipulates that where the employer fails to pay any amount due to the contractor by the final date for payment then the employer shall also pay simple interest to the contractor at the rate of 5% over the base rate of the Bank of England which is current at the date the payment became overdue. See Section 9.6 for a more detailed discussion of interest.

In certain circumstances the employer may be entitled to deduct sums from the contract price before making payment. Where the contractor has delayed completion of the works then the employer may be entitled to deduct liquidated damages, see Section 5.9. The employer may also have an obligation to deduct income tax from payments to the contractor. Clause 31 of JCT 98 makes specific provision for such deductions. Employers may also have rights under the contract to make deductions in respect of costs they have incurred. For example, clause 21.1.3 of JCT 98 provides that if the contractor fails to take out insurance to cover death or personal injury to third parties then the employer can take out such insurance and deduct the premiums from any payments otherwise due to the contractor.

### 7.2.2 Payment for contractual variations

As a result of the magnitude and complexity of most building contracts, there will inevitably arise during the carrying out of the works a need to add, omit or vary some of the work to be executed. For this reason most building contracts, and certainly all the common standard forms of building contract, contain detailed provisions to govern the instruction, execution and payment of variations, see Section 3.5 for a discussion of variations.

A building contract can provide that the contractor will be entitled to no additional payment in respect of such variations. As the contractor is obviously unlikely to agree to such provisions, they are uncommon. It is more normal for provision to be made that the contractor will be entitled to additional payment for variations. Clause 30 of JCT 98 specifically provides that the contract sum should be adjusted by the value of any variations. Nevertheless, in order to ensure that the contractor is entitled to recover additional payment, it is important that any variations are instructed in accordance with the terms of the contract. If the variations are not so instructed, or if the contractors cannot prove that they were so instructed, then they may have difficulty recovering payment, see for example *Robertson* v. *Jarvie* (1907).

In order to try and avoid disputes as to whether variations were or were not instructed, many building contracts provide that all variations should be instructed in writing. In such cases, the contractor may, in the absence of a written instruction, be unable to recover payment under the terms of the contract, see for example *Brown* v. *Lord Rollo and Others* (1832) and *Holland Hannen & Cubitts (Northern) Ltd* v. *Welsh Health Technical Services Organisation and Others* (1981). Accordingly, where the contractor is instructed to execute a variation by a mechanism other than that provided for in the contract, the advisable course of action for the contractor is to refuse to carry out the work until the correct procedure has been followed. Otherwise the contractors run the risk that they will be unable to recover payment.

Disputes often arise as to whether an instruction to the contractor is or is not a variation under the contract. For example, a dispute may arise as to whether the contractor is obliged to execute a particular item of work as part of the original contract works or whether that particular item of work is a variation. In such a case, by refusing to carry out the work because of the absence of an instructed variation, the contractor runs the risk of being in breach of contract if it is ultimately determined that the work did form part of the original contract works. In practice, the contractor often simply carries out the work and makes a claim for payment despite the lack of an appropriate variation order.

Similar difficulties can arise where the contractor is instructed to rectify or alter work which it is alleged is not in accordance with the contract. If the work was in accordance with the contract then the contractor will have incurred the cost of carrying out further work for which they have no appropriate variation order. In such cases the courts have, in certain circumstances, held that the contractor is entitled to payment where a variation order should have been issued, but was improperly withheld, see *Brodie* v. *Corporation of Cardiff* (1919). In other cases the courts have construed instructions not purporting to be variations to be in fact, variations, see *Shanks & McEwan (Contractors) Ltd* v. *Strathclyde Regional Council* (1995).

Where the variation is made for the contractors benefit, however, the Courts have not been willing to construe instructions as variations entitling the contractor to additional payment, see *The Tharsis Sulphur and Copper Company Ltd* v. *McElroy & Sons* (1878).

### 7.2.3 Payment for variations under JCT 98

Where a building contract provides a mechanism to regulate variations to the works, it is common and, it is submitted, prudent for the contract also to contain provisions to govern how any such variations are to be valued. The method of valuation can take a number of different forms and indeed many of the standard forms of building contract contain more than one method of valuation. The relevant provisions in JCT 98 are to be found in clauses 13 and 13A which provide a number of possible alternatives for valuation.

In the first place, it is provided that the employer and contractor can agree how a variation is to be valued. Unfortunately, this rarely happens in practice as many variations require immediate compliance leaving little time for agreement of a price in advance between the contractor and employer. It is, however, a useful mechanism, particularly in relation to major variations. In order to try and facilitate the agreement of a price in advance, the architect when instructing a variation can stipulate in the instruction that clause 13A is to apply and, accordingly, the provisions of that clause will regulate the valuation of the variation unless the contractor disagrees in writing to the application of the clause within seven days of receipt of the instruction. Under clause 13A, the contractor is required, on receipt of the instruction, to submit a quotation to the quantity surveyor. Clause 13A specifies in detail what the quotation should contain. The quotation can then be accepted by the employer and the acceptance needs to be confirmed by the architect. Clause 13A sets out the time scales within which this should take place. Until the quotation is accepted the variation is not executed by the contractor. If the quotation is not accepted by the employer then the architect should either instruct that the variation is not to be carried out or that it is to be carried out but valued in accordance with the provisions of clause 13.

The purpose of clause 13A is to deal with variations which the employer may require but for which they wish to know the price prior to confirming that the work is to be carried out. Clause 13, on the other hand, provides a mechanism for valuing variations which the contractor is contractually bound to carry out when instructed and for which the employer is contractually bound to pay. Accordingly, clause 13 is used to value all instructed variations other than those to which clause 13A applies or those where the employer and contractor have agreed a price in advance.

Clause 13 provides two distinct methods for valuing variations, namely, 'Alternative A: Contractor's Price Statement' and 'Alternative B'.

Alternative A is for the contractors to adopt and it allows them to submit a price statement to the quantity surveyor on receipt of an instruction stating their price for carrying out the variation. Clause 13.4.1.2 specifies what the price statement should include and sets out the procedure to be followed and the time scales within which the price statement should be submitted and within which it can be accepted in whole or in part. The price statement should be based on the detailed valuation rules found in clause 13.5 in order that it can be analysed by the quantity surveyor and thereafter accepted following consultation with the architect. To the extent that the price statement is not agreed then an amended price statement should be prepared by the quantity surveyor and supplied to the contractor for acceptance. If the amended price statement is not accepted by the contractor then either party may refer the price statement, or the amended price statement, to adjudication as a dispute. In the meantime, however, the contractor should comply with the instruction (in contrast to the position under clause 13A).

Alternative B applies where the contractor does not implement Alternative A, or where it has been implemented but where the price statement or amended price statement have not been accepted and neither party has referred the dispute to adjudication. Under Alternative B the quantity surveyor must value the variation in accordance with the valuation rules found in clause 13.5. Clause 13.5 provides that where the variation requires the execution of additional or substituted work which can properly be valued by measurement then it shall be valued in accordance with specific rules.

- Where the additional or substituted work is of similar character to, is executed under similar conditions as, and does not significantly change the quantity of, work set out in the contract bills then the rates and prices for the works set out in the contract bills shall determine the valuation.
- Where the additional or substituted work is of similar character to work set out in the contract bills but is not executed under similar conditions and/or significantly changes the quantity thereof, then the rates and prices for the work so set out shall be the basis for determining the valuation and the valuation shall include a fair allowance for such difference in conditions and/or quantity.
- Where the additional or substituted work is not of similar character to work set out in the contract bills then the work shall be valued at fair rates and prices.
- To the extent that the variation relates to the omission of work set out in the contract bills then the rates and prices for such work therein set out shall determine the valuation of the work omitted.
- To the extent that the variation requires the execution of additional or substituted work which cannot properly be valued by measurement then clause 13 provides for valuation on a cost basis.

If the valuation of a variation cannot reasonably be effected using any of the foregoing methods then clause 13.5.7 provides that a fair valuation shall be made.

Regardless of whether the valuation is calculated under Alternative A, Alternative B or clause 13A, clause 13.7 provides that the valuation will be effected by addition to or deduction from the contract sum. The various methods of valuation also allow for valuation of any loss and expense arising from the variation (see section 7.3), and also allow for adjustment to the time for completion of the works (see Section 5.5).

As can be seen, the provisions for valuing variations in JCT 98 are extremely detailed and have evolved over a number of years. Similar provisions are to be found in a number of the other standard forms of building and engineering contracts although it is open to the parties at the time of contracting to agree any method they choose for valuing variations. Where the contract does not provide a mechanism then variations will require to be valued on the basis of *quantum meruit*, see section 7.4.

### 7.2.4 Fluctuation in cost

In concluding a price for any building contract there is a risk to both the contractor and the employer that the costs involved in constructing the building can fluctuate dramatically due to changes in economic factors entirely outwith the control of either of the parties. Such economic factors can include inflation, which can affect the price of both labour and materials, and also changes in tax legislation.

The risk to the employer is that costs decrease and they therefore have to pay the contractor far more than the building actually cost to construct. On the other hand this may not be a substantial risk as the employer will have budgeted for the contract price prior to concluding the contract.

The risk to the contractor is that the costs increase resulting in either a diminution in their profit or, more worryingly, the building costing more to construct than the price contracted for. It is submitted that such an eventuality is not in the interest of either the contractor or the employer as it can result in the contractor having difficulties completing the building and, in some instances, may give rise to the contractor trying to cut corners to minimise costs.

Fluctuations in cost are a particularly serious risk where the contract price is substantial or the contract period particularly lengthy. Without any provision in the contract to deal with such fluctuations in cost both parties would be bound by the price agreed. Trying to assess the potential effect of fluctuations can be extremely speculative and, indeed, it could result in buildings becoming more expensive to construct as contractors increase their price to minimise their exposure to cost fluctuations.

In an effort to try and minimise the risk of fluctuation in cost, many of the standard forms of building contract provide a mechanism for adjusting the contract price to take account of increases or decreases in cost during the contract period. Such provisions can be found in clauses 38, 39 and 40 of JCT 98. These clauses provide three alternatives for dealing with fluctuations. The parties should choose which alternative is to be employed at the time of entering into the contract and the choice should be inserted in the Abstract of Conditions found in Appendix II of the Scottish Building Contract.

If no choice is made then clause 37 and the Abstract of Conditions stipulate that the alternative contained in clause 38 is to apply. It should be noted that the fluctuation in cost provisions do not apply to variations where a price has been agreed in advance under clause 13A, see Section 7.2.3 above. Of the three alternatives:

- clause 38 provides for adjustment in the contract price as a result of changes in contributions, levies and taxes,
- clause 39 provides for adjustment in the contract price as a result of fluctuations in the cost of labour and materials and as a result of tax changes, and
- clause 40 provides for adjustment using a price adjustment formula.

If either party is to rely on the fluctuation provisions then it is important that they fully comply with any pre-requisites contained within the relevant clause, see *John Laing Construction Ltd* v. *County and District Properties Ltd* (1982).

## 7.3 Direct loss and expense

In building contracts, it is not uncommon for the contractor's progress with the contract works to be affected by events that are within the control of the employer, or the architect. Clause 26 of JCT 98 contains a detailed mechanism that, in certain circumstances, entitles the contractor to recover direct loss and/or expense where the regular progress of the works has been delayed or where there has been a deferment of giving possession of the site to the contractor, under clause 23.1.2. Clause 26 is not exhaustive of the contractor's rights in such circumstances. Clause 26.6 provides that the provisions of clause 26 are without prejudice to any other rights and remedies which the contractor may possess. There is no direct connection between the extension of time provisions in clause 25 of JCT 98 and the provisions for recovery of direct loss and/or expenses in clause 26. The two clauses have distinct and separate purposes. Support for this proposition can be found in *Methodist Homes Housing Association Ltd* v. *Scott & McIntosh* (1997).

As soon as contractors become aware, or should reasonably have become aware, that the regular progress of the works has been affected, they need to make written application to the architect stating that they have incurred or are likely to incur direct loss and/or expense for which they would not be reimbursed by payment under any other provision of the contract. Direct loss and/or expense is that which flows naturally in the usual course of things, see *F G Minter Ltd* v. *Welsh Health Technical Services Organisation* (1980), following *Saint Line Ltd* v. *Richardsons Westgarth & Co Ltd* (1940).

A particular item of damage or loss or expense is to be regarded as 'direct' if it falls within the first rule in *Hadley* v. *Baxendale* (1854), which is considered in detail below at Section 9.4.2. Direct loss and/or expense flows naturally from the matters provided for in clause 26 if there is no other intervening cause, see *Croudace Construction Ltd* v. *Cawoods Concrete Products Ltd* (1978). Contractors must actually incur direct loss and/or expense before they can recover under clause 26, albeit that they are obliged to give notice as and when it becomes apparent that they are likely to incur direct loss and/or expense.

A list of matters in respect of which direct loss and/or expense can be claimed is contained in clause 26.2. The matters are:

- a failure timeously to release the information referred to in an information release schedule or to provide further drawings or details;
- the opening up for inspection of work under clause 8.3, unless the inspection shows that the work was not in accordance with the contract;
- any discrepancy in or diversions between the contract drawings and/or the contract bills and/or the numbered documents;
- the execution of work not forming part of the contract by the employer himself or by persons employed or otherwise engaged by the employer or their failure to execute such work (this does not include nominated sub-contractors);
- the supply by the employers of materials and goods which they have agreed to provide for the contract works or their failure so to supply;
- architect's instructions under clause 23.2 issued in regard to the post-ponement of any work to be executed under the provisions of the contract;
- failure by the employer to give in due time ingress to or egress from the site;
- the issue of instructions requiring a variation;
- the execution of work for which an approximate quantity is included in the contract bills which is not a reasonably accurate forecast of the quantity of work required;
- compliance or non-compliance with duties in relation to the CDM Regulations; and
- suspension by the contractor of the performance of his obligations under the contract to the employer under clause 30.1.4 (failure to pay a sum due under an interim certificate) providing the suspension was not frivolous or vexatious.

The contractor's written application must be framed with sufficient particularity to enable the architect to do what he is required to do under the contract. The application must be made within a reasonable time. It must not be made so late that the architect can no longer form a competent opinion of the matters on which he is required to satisfy himself that the contractor has suffered the loss or expense claimed. In considering whether the contractor has acted reasonably, it must be borne in mind that the architect is not a stranger to the work and that it was always open to the architect to call for further information either before or in the course of investigating a claim, see *London Borough of Merton* v. *Stanley Hugh Leach Ltd* (1985).

Having received the contractor's written application the architect needs to satisfy himself as to whether or not direct loss and/or expense has been incurred or is likely to be incurred. Assuming he is so satisfied, then the architect from time to time shall ascertain, or shall instruct the quantity surveyor to ascertain, the amount of such loss and/or expense which has been or is being incurred by the contractor. The architect can request information to enable him to form an opinion, but only the architect can decide upon the validity of a claim, see *John Laing Construction Ltd* v. *County and District Properties Ltd* (1982). Either the architect or the quantity surveyor can request details of alleged loss and/or expense for the purposes of ascertaining the amount thereof due to the contractor. Under clause 30.9.1.4 a final certificate is conclusive evidence that the reimbursement of direct loss and/or expense, if any, to the contractor pursuant to clause 26.1 is in final settlement of all and any claims which the contractor has or may have arising out of the occurrence of any of the matters referred to in clause 26.2, whether such claim be for breach of contract, duty of care, statutory duty or otherwise. Despite the provisions of clause 26.6, the final certificate is conclusive as to loss and expense in respect of any of the matters described in clause 26.2, notwithstanding the legal basis upon which payment arising out of such matters is sought.

## 7.4 *Quantum meruit*

It is a general principle of Scots law that the recipient of services in terms of a contract is under an implied obligation to pay for the services. As indicated previously the final contract price or a mechanism for ascertaining the final contract price should be agreed at the time of entering into a building contract. Similarly it is prudent to agree the basis on which any variations or additions will be priced. In practice, this sometimes does not happen. In such circumstances, the contractor may still be entitled to payment *quantum meruit*, that is payment of a reasonable sum for the work carried out by them.

In order for a claim for *quantum meruit* to succeed there must be a contract between the parties, see for example *Alexander Hall & Son (Builders) Ltd* v. *Strathclyde Regional Council* (1989). There is no scope for a *quantum meruit* claim if the contract between the parties contains an agreed contract price or a mechanism for ascertaining the contract price, see for example *Interbild Components Ltd* v. *Fife Regional Council* (1988). As has been indicated previously, however, agreement of the price, or a mechanism for ascertaining the price, is an essential term of any building contract and, in the absence of such agreement, there may be no binding contract between the parties.

In certain circumstances, however, where work has been carried out by agreement and only the price has not been agreed, the courts have been prepared to hold that an implied contract exists between the parties. See for example *Avintair Ltd* v. *Ryder Airline Services Ltd* (1994) in which services were performed by one party but no price was agreed. The court held that in those circumstances the law would imply, from the parties' conduct, a contract that a reasonable sum be paid and that the appropriate claim in those circumstances was on implied contract on the principle of *quantum meruit*.

Before the law can imply such a contract, however, it appears that the services must already have been performed by the party seeking payment. See for example *British Bank for Foreign Trade Ltd* v. *Novinex* (1949) in which Lord Denning stated:

> 'In the ordinary way, if there is an arrangement to supply goods at a price "to be agreed", or to perform services on terms "to be agreed", then although, while the matter is still executory, there may be no binding contract, nevertheless, if it is executed on one side, that is, if the one does his part without having come to an agreement as to the price or the terms, then the law will say that there is necessarily implied, from the conduct of the parties, a contract that, in default of agreement, a reasonable sum is to be paid.'

In addition to the contract price, the claim *quantum meruit* can also apply to additional work instructed where no price has been agreed, see for example, *Taylor* v. *Andrews-Weatherfoil Ltd* (1991). If, however, the additional works have not been instructed then the claim will fail, see for example *T & R Aitken* v. *Gardiner* (1958). *Quantum meruit* can also apply where a price has been agreed but it becomes inapplicable through the passage of time. See for example *Constable Hart & Co Ltd* v. *Peter Lind & Co Ltd* (1978) in which a price was agreed which was fixed until a particular date. The contract was delayed through no fault on the part of the sub-contractor and the court held that work carried out after the agreed date had been carried out at reasonable rates.

In certain circumstances, even where a price has been agreed, it may be possible for contractors to put forward a claim that they can ignore the contract price and demand payment on the basis of *quantum meruit*. In certain circumstances, it may be possible for the contractor to show that, as a result of breaches of contract on the part of the employer, the contract works have altered so dramatically from those contracted for that the contract price is no longer applicable and a new contract term should be implied that they be paid on the basis of *quantum meruit*, see *Lodder* v. *Slowey* (1904). Before contractors can put forward a claim on the basis of *quantum meruit*, however, it appears that they must rescind the contract in order to make it clear that they no longer consider themselves bound by the original contract price, see *ERDC Construction Ltd* v. *H M Love & Co* (1995), *Boyd & Forrest* v. *Glasgow & South Western Railway Company* (1915) and *Smellie* v. *Caledonian Railway Company* (1916). In order to allow the contractor to rescind any breach by the employer will accordingly have to be material. In the event that the breach is not material, or should the contractor choose not to rescind, then the contractor's remedies will be limited to payment of the contract price for the work executed together with damages for breach of contract. In *Morrison-Knudsen Co Inc* v. *British Columbia Hydro & Power Authority* (1978) it was stated that:

> 'It is well established law that a plaintiff's remedies for a defendant's default under a contract between them are limited to those provided in the contract or which may be awarded for breaches of the contract for so long as the contract remains open and available to the parties. To enable the court to award compensation by *quantum meruit* the Respondents must show that [the contract] has been rescinded or discharged and that mutual obligations thereunder have ceased to exist. While it continues to exist the obligation of [the plaintiff] and the rights of the Respondents are limited by its terms.'

This statement was cited with approval in *ERDC Construction Ltd* in which the court indicated that a party faced with a breach of contract had to elect between affirming the contract and holding the other party to the performance of its obligations or, alternatively, rescinding the contract and suing at once for damages or *quantum meruit* for performance to the date of rescission. The court indicated that the election must be made promptly and communicated to the employer and once made would be binding on the parties and could not be changed. If the contractor simply continues to carry out the contract works then he waives his right to claim payment *quantum meruit*, see *Smellie*.

Another situation in which it may be possible to ignore the contract price and seek payment on the basis of *quantum meruit* is where the nature of the work carried out is altered fundamentally from that which the contractor originally contracted to carry out. As a result the works may become more difficult and more expensive. In such circumstances it may be open to the contractor to maintain that the original contract has been frustrated by the fundamental alterations and that they are entitled to maintain a claim based on *quantum meruit*, see for example *Head Wrightson Aluminium Ltd* v. *Aberdeen Harbour Commissioners* (1958) and *Smail* v. *Potts* (1847).

The alteration to the work may not amount to breach of contract as the employer may, for example, have power to instruct variations in terms of the contract. Similarly, the contractor may suggest variations which are approved by the employer, see *Mercer* v. *Wright* (1953). It appears that if a contractor wishes to claim for payment *quantum meruit* he may have to advise the employer at the time when the difficulty becomes apparent, see for example *Mackay* v. *Lord Advocate* (1914). It is submitted that this is correct as it offers the employer an opportunity to choose between proceeding with the contract works on the basis of payment *quantum meruit* or halting the work because of the frustration.

If, however, the works have become more difficult or more expensive because of matters which existed at the time of contracting, but which the contractor did not foresee, then the contract will not be frustrated and the contractor will have no entitlement to payment *quantum meruit*, see *Davis Contractors Ltd* v. *Fareham UDC* (1956).

Where a contractor is entitled to make a claim for payment *quantum meruit* then they are entitled to be paid at ordinary or market rates, or where no such rates are available they are entitled to be paid a reasonable rate, see

*Avintair Ltd.* The party seeking payment *quantum meruit* is entitled to lead evidence to prove what would be a reasonable rate in the circumstances, see *Wilson* v. *Gordon* (1828). It appears that a building contractor can include in their rate elements for work carried out, material supplied, overhead costs and reasonable profit, see for example *Monk Construction Ltd* v. *Norwich Union Life Assurance Society* (1992). In addition to proving that the rate charged is reasonable, the party claiming payment will also have to prove to the court that the amount of time spent carrying out the work was reasonable, see *Scottish Motor Traction Co* v. *Murphy* (1949).

## 7.5 *Quantum lucratus*

Disputes sometimes arise in a situation where a person has in good faith carried out works to another's land or property where there is no contract between the parties. In such a situation the party carrying out the work cannot claim payment in terms of the contract or payment *quantum meruit*. There may, however, be an entitlement to payment on a *quantum lucratus* basis, that is seeking to recover the value of the land or property owner's enrichment. See for example *Newton* v. *Newton* (1925). *Quantum lucratus* is a branch of the law of recompense and is an equitable remedy requiring an owner of land to pay for works carried out by another on their land as a result of which they are enriched. *Quantum lucratus* does not apply to the situation where a third party is enriched by work carried out by the owner of land or property to that land or property, see *Edinburgh and District Tramways Company Ltd* v. *Courtenay* (1909). It may, however, apply in the situation where work is carried out to common property, see *Stark's Trustees* v. *Cooper's Trustees* (1900).

Where work is carried out to land or property without the owner's permission or agreement then the owner can insist that the works are removed. If the owner does not do so then the principle of *quantum lucratus* deems him to have accepted the benefit and requires him to pay for that benefit. If, however, the person who has carried out the works has done so in bad faith then he will have no claim based on *quantum lucratus*, see *Barbour* v. *Halliday* (1840) and *Duke of Hamilton* v. *Johnston* (1877).

If works are carried out in terms of a contract between the parties there is no scope for a *quantum lucratus* claim, so long as the contract remains applicable, see *Thomson* v. *Pratt* (1962) It appears that where the contract becomes inapplicable because it has been determined by the employer who retains the benefit of any works already carried out then the contractor may be entitled to a *quantum lucratus* claim, see *Alexander Graham & Co* v. *United Turkey Red Company Ltd* (1922), *NV Devos Gebroeder* v. *Sunderland Sportswear Ltd* (1990) and *R & J Scott* v. *Gerrard* (1916). A *quantum lucratus* claim cannot be sustained if the contract makes provision for payment on its determination, see for example clause 27 of JCT 98.

The courts have had difficulty laying down a general definition for *quantum lucratus* and have indicated that each case requires to be looked at in its own particular circumstances, see *Edinburgh and District Tramways Company Ltd, Varney (Scotland) Ltd* v. *Burgh of Lanark* (1976) and *Lawrence Building Co Ltd* v. *Lanarkshire County Council* (1979). It appears, however, that some essential features must exist in order for a claim to succeed. In both *Varney (Scotland) Ltd* and *Lawrence Building Co Ltd* it was held that for a claim to succeed:

- the pursuers must have incurred a loss although the cost of carrying out the works will suffice in this respect;
- the pursuers must not have intended to make a gift to the defenders; and
- there must be a quantifiable benefit to the defenders who are thereby *lucrati.*

Accordingly, if the pursuers carry out the work for their own benefit then they will not be entitled to claim *quantum lucratus*, see *Edinburgh and District Tramways Company Ltd* and *Rankin* v. *Wither* (1886). Some incidental benefit will not, however, bar a claim. It also appears that a claim for *quantum lucratus* cannot succeed where the pursuer has any other legal remedy available, see *Stewart* v. *Stewart* (1878).

In *Varney (Scotland) Ltd* it was stated that:

> 'Recompense is an equitable doctrine. That being so, it becomes a sort of court of last resort, recourse to which can only be made when no other legal remedy is or has been available. If a legal remedy is available at the time when the action which gave rise to the claim for recompense has to be taken, then normally that legal remedy should be pursued to the exclusion of a claim for recompense.'

Some authorities have indicated that a claim for *quantum lucratus* can only succeed where there has been an error or mistake of fact on the part of the person making the claim, see *Rankin* v. *Wither* (1886), *Buchanan* v. *Stewart* (1874) and *Gray* v. *Johnston* (1928). This can be contrasted, however, with the comments of Lord Justice Clerk Alness in *Gray* who stated that he did not think error was essential in all cases for a claim for *quantum lucratus* to succeed. Similarly in *Varney (Scotland) Ltd* the court indicated that error may found a claim for *quantum lucratus* but that the absence of an error or mistake of fact will not invalidate a claim if the other circumstances justify its imposition.

## 7.6 *Contractual retention*

In the standard forms of building contract it is common to find a mechanism whereby the employer can deduct an amount from any payment otherwise due to the contractor by way of contractual retention. The purpose of the retention is to allow the employer to retain a proportion of any payment due in respect of work already carried out as security against the risk of any failure by contractors to complete their obligations under the contract, including the making good of defects. Once the contractor has completed all his obligations in terms of the contract then the retention is released.

Accordingly, the employer can use the retention as a lever to ensure the contractor completes the works or, alternatively, as a fund to pay for completion of the works in the event that the contractor does not fulfil his obligations. This can be particularly important in the event of the contractor's insolvency where, without any retention, the employer might simply be left with an unsecured claim against the contractor for breach of contract. See for example *Asphaltic Limestone Concrete Co Ltd and Another* v. *Corporation of the City of Glasgow* (1907).

The standard forms of building and engineering contracts normally contain detailed rules governing both the deduction of contractual retention and its release. Clause 30.2 of JCT 98 provides that the amount stated as due in an interim certificate shall be the gross valuation of the work carried out by the contractor, less any retention as provided for in clause 30.4. Clause 30.4 provides that the retention which the employer may deduct and retain shall be a percentage of the total amount included in any interim certificate. The percentage is stipulated to be 5% unless a lower rate is agreed between the parties and inserted in Appendix II of the Scottish Building Contract at the time of entering into the contract.

The retention percentage may be deducted from the amount certified in any interim certificate insofar as the amount certified relates to work which has not reached practical completion. Where the work has reached practical completion but no certificate of completion of making good defects has been issued then the employer may only deduct half the retention percentage. In practice this operates on the basis that the employer deducts the whole retention percentage from amounts included in interim certificates issued to the contractor. Half the retention is then released on practical completion of the works with the remaining half being released on the issue of a certificate of completion of making good defects.

Clause 30.5 of JCT 98 stipulates that the employer's interest in the retention is fiduciary as trustee for the contractor. At the date of each interim certificate the architect has to prepare or instruct the quantity surveyor to prepare a statement specifying the retention deducted in arriving at the amount stated as due in the interim certificate and this statement is issued to the contractor. Thereafter, the contractor can request the employer, at the date of payment under each interim certificate, to place the retention to be deducted in a separate bank account and certify that this has been so done. It appears that the contractor can also make this request at a later date if he has not done so at the date of payment. See *J F Finnegan Ltd* v. *Ford Sellar Morris Developments Ltd* (1991). The employer is entitled to any interest accruing on the retention while it remains in this separate account.

The objective of clause 30.5 is to provide a mechanism whereby the retention deducted by the employer is to be held in trust on behalf of the contractor. See *Wates Construction (London) Ltd* v. *Franthom Property Ltd* (1991). This is to afford the contractor a degree of protection in the event of the insolvency of the employer. If it is not placed in a separate account then it appears that the contractor will have no protection, see *Mac-Jordan Construction Ltd* v. *Brookmount Erostin Ltd* (1991). If the employer fails to put the money in a separate account then the contractor's remedy would be an action for specific implement (see Section 9.3 below). Given the time it may take to conclude such an action, this remedy may be of little practical assistance. This is particularly so in cases where the employer disputes that the contractor is entitled to the retention because the employer has other claims which he wishes to meet out of the contractual retention, for example liquidated damages, see *Henry Boot Building Ltd* v. *The Croydon Hotel & Leisure Co Ltd* (1985) and *GPT Realisations Ltd (in Administrative Receivership and in Liquidation)* v. *Panatown Ltd* (1992). Contrast, however, the decision in *Concorde Construction Co Ltd* v. *Cogan Co Ltd* (1984).

Unfortunately, it appears that under Scots Law the terms of clause 30.5 are, themselves, insufficient to create a trust without other actings on the part of the employer, see *Clark Taylor & Co Ltd* v. *Quality Site Development (Edinburgh) Ltd* (1981) and *Balfour Beatty Ltd* v. *Britannia Life* (1997). Accordingly, in Scotland, if an employer becomes insolvent then, in respect of the payment of retention which has been deducted, the contractor may find themselves in no better a position than other ordinary creditors. This may differ from the position under English Law where a trust can be established by less formal means. A detailed examination of the law of creation of trusts is beyond the scope of this book.

A practice has grown up whereby, in order to receive payment of the full amount of interim certificates and the final account, the contractor will often provide the employer with a bond equivalent to the amount which would otherwise have been retained by the employer until full satisfaction of the works by the contractor, including remedying defects. These bonds are either put in place from the commencement of the works or are put in place at the time of practical completion in respect of the remaining half of the retention fund which would otherwise not be payable until after the certification of the making good of defects.

# Chapter 8
# Ending a Building Contract

## 8.1 *Introduction*

Having examined what is required to constitute a building contract in Scotland, and the rights and obligations arising, it is important to establish when, and in what circumstances, a building contract and the obligations arising from it will be brought to an end. The law of Scotland contains a number of general rules which relate to the extinction of contractual obligations, and these apply equally to the extinction of the obligations under a building contract.

Certain of the methods by which an obligation may be extinguished are of general application rather than being peculiar to building contracts. For example, an obligation may be extinguished by acceptilation, where the creditor discharges his right without payment or performance. An obligation for the payment of money may also be extinguished by confusion, where the same person becomes creditor and debtor in an obligation. This does not apply where there are continuing rights and obligations beyond the payment of money, and thus, were the employer under a building contract to take-over the contractor (or vice-versa) during the currency of the contract, the doctrine of confusion would not apply. A detailed examination of these doctrines is beyond the scope of this book.

It must be borne in mind that, at any time during the currency of a contract, it is open to parties to enter into an agreement whereby their respective obligations are extinguished. Building contracts habitually contain detailed mechanisms whereby one or other of the parties, or both of them, is entitled to determine the contract.

It is the intention of this chapter to examine certain methods of extinction of obligations that are of particular significance to building contracts. Certain others, such as payment (Chapter 7) and novation (Chapter 11) are dealt with elsewhere in this book.

## 8.2 *Frustration and impossibility*

Frustration occurs whenever, without fault on the part of either party, intervening circumstances have rendered a contract incapable of being performed, or so altered the conditions that, if there were to be performance, it would, in essence, be performance of a different contract, see *Davis Contractors Ltd* v. *Fareham UDC* (1956). In judging whether or not a contract has been frustrated the contract must be viewed as a whole. The question to be considered is whether the purpose of the contract, as gathered from its terms, has been defeated, see *James B Fraser & Co Ltd* v. *Denny, Mott & Dickson Ltd* (1945). It follows that if parties had regard to the possibility of a particular event, and made provision for it in their contract, the occurrence of such an event cannot have the effect of frustrating the contract, see *Cricklewood Property & Investment Trust Ltd* v. *Leighton's Investments Trust Ltd* (1945). If the contract does not contemplate the intervening circumstances it will be frustrated.

Whether or not a contract has been frustrated will, in each case, be a question of fact to be decided upon the true construction of the terms of the contract, read in the light of the nature of the contract and of the relevant surrounding circumstances when the contract was made, see *Head Wrightson Aluminium Ltd* v. *Aberdeen Harbour Commissioners* (1958).

In the leading case of *Davis Contractors Ltd*, the House of Lords held that a contract which had been scheduled to take eight months, and was said to be subject to there being adequate supplies of labour available as and when required, but which took 22 months to complete due to unanticipated shortages of labour and materials, had not been frustrated. The qualification as to the availability of adequate supplies of labour was contained in a letter which accompanied the contractors' tender. That letter was held not to form part of the contract and the contractors had to bear the additional costs.

Nevertheless, there may be circumstances where modifications, which necessarily and fundamentally alter the whole design of a project, frustrate the original contract and entitle the contractor to a claim based upon *quantum meruit*. This is considered more fully in Section 7.4. The absence of an intimation by a contractor that he is proceeding upon a *quantum meruit* basis may be an important element in deciding whether there has, in fact, been frustration.

Another example of frustration is where the performance of a contract is dependent upon a certain thing existing and that thing is either destroyed or is so fundamentally altered that the contract cannot be performed. This is known as *rei interitus*. If this occurs prior to the contractor taking possession of the site then neither party will have a claim against the other. If it occurs when building works are underway the contractor has a claim for the work carried out and the materials supplied. By the doctrine of accession, property in the building passes to the owner of the ground upon which it is erected and the contractor's entitlement to payment arises under the principle of *res perit domino* (a thing perishes to its owner). A contractor may not be entitled to payment if the work carried out is so defective that the employer would have a defence to an action raised against him. If payments have been made in advance and the contract is subsequently frustrated the payments made can be recovered under the doctrine known as *condictio causa data non secuta* (a claim that the consideration has failed of its purpose), see *Cantiere San Rocco, SA* v. *Clyde Shipbuilding and Engineering Co Ltd* (1923).

Where the contract is frustrated it is more accurately parties' rights and obligations as to future performance under the contract that are frustrated.

In the context of building contracts this distinction is important as even after frustration certain clauses, most notably arbitration clauses, may continue to be enforceable, see *Heyman and Another* v. *Darwins Ltd* (1942).

## 8.3 *Force majeure*

### 8.3.1 General

As frustration cannot apply where the parties to a contract have had regard to the possibility of a particular event and made provision for it in their contract, difficulties of interpretation may arise in determining whether frustration has occurred. To address this problem many contracts expressly provide for events that might ordinarily be sufficient to frustrate the contract. Such clauses are known as *force majeure* clauses.

The term *force majeure* is believed to emanate from France, and in particular the Code of Napoleon. It has no particular technical meaning in Scotland. The term covers events beyond the control of the party to the contract who seeks to rely upon the clause such as war, epidemics and strikes. It is also said to encompass any direct legislative or administrative interference, see *Lebeaupin* v. *Richard Crispin and Co* (1920).

By its very nature a *force majeure* clause will have to be read carefully in conjunction with the remaining terms of the contract to establish, precisely, its scope.

### 8.3.2 *Force majeure* under JCT 98

By virtue of clause 25.4.1 of JCT 98, *force majeure* constitutes a relevant event which may give rise to an extension of time. Civil commotion and the use or threat of terrorism are separate relevant events under clause 25.4 although they might otherwise, in any event, fall within the ambit of a *force majeure* clause.

Clause 28A.1.1.1 of JCT 98 provides that, if *force majeure* causes the carrying out of the contract works to be suspended for a continuous period of time specified in the contract, either the employer or the contractor is entitled to determine the employment of the contractor. This is considered further in section 8.4.4 below.

Clause 36.4.3 provides that the delivery programme agreed between the contractor and the nominated supplier may be varied by reason of *force majeure*.

## 8.4 *Determination*

### 8.4.1 Contractual provisions

The majority of building contracts contain express provisions regulating the rights of either or both of the parties in defined circumstances to determine the contract, or bring it to an end. In exercising such rights parties should exercise extreme caution and ensure that the determination procedure laid down in the contract is strictly adhered to, see *Muir Construction Ltd* v. *Hambly Ltd* (1990). Where the contract has already been brought to an end it may not be determined, see *W Hanson (Harrow) Ltd* v. *Rapid Civil Engineering Ltd and Another* (1987). A party who purportedly operates a determination clause in circumstances where they are not entitled to do so may be treated as having repudiated the contract, see *Architectural Installation Services Ltd* v. *James Gibbons Windows Ltd* (1989).

By purporting to determine the contract a party is clearly indicating that they are not going to perform in the future. Unless they are permitted to do that under the contract such an intention constitutes a repudiation. In such circumstances the other party to the contract is entitled to accept the repudiation, rescind the contract and seek damages. The concepts of repudiation and recission are considered below at section 8.5. Such a state of affairs may prove welcome to the recipient of the purported determination notice, particularly if that party was finding it difficult to perform in the first place!

Assuming the contract is properly determined, what is the effect of that determination? The contract as a whole is not terminated. While many of the obligations under the contract, including what might conveniently be termed the principal obligations (e.g. the obligation of the contractor to execute the contract works), will no longer be enforceable, the remaining obligations are fundamentally altered but continue to have effect, see *MacJordan Construction Ltd* v. *Brookmount Erostin Ltd* (1991). Ordinarily a determination clause will also provide for the respective rights and duties of the parties in the event of such a clause being operated. By their nature, those provisions are intended to operate upon the contract being determined.

An arbitration clause will continue to be operative, notwithstanding the fact that the contract has been determined, see *R & J Scott* v. *Gerrard* (1916). Where the contract contains provisions that deal with an assessment of the sums due to or by either party and an accounting therefore on determination, the courts in Scotland have held that a claim based upon an alleged breach of contract is irrelevant. The correct way to proceed is to claim for payment based upon the contractual provisions, see *Muir Construction Ltd.*

In this section we will examine the determination provisions under the Scottish Building Contract. The determination provisions can conveniently be considered under three headings, namely:

- the circumstances in which there can be determination by the employer,
- the circumstances in which there can be determination by the contractor, and
- the circumstances in which there can be determination by either party.

In each case, it is the employment of the contractor under the contract that is determined, not the contract itself.

### 8.4.2 Determination by employer under JCT 98

Clause 27 of JCT 98, which provides for determination by the employer, is deleted and replaced by a new clause in the Scottish Building Contract. The new clause applies equally whether the Private or Local Authorities Edition of JCT 98 is used, and entitles the employer to determine the employment of the contractor if the contractor continues a 'specified default' for 14 days after receiving from the architect (the contract administrator in Local Authorities contracts) a notice specifying that default.

The 'specified defaults' relied upon must arise before the date of practical completion and be one of those set out in clause 27.2.1. These are:

(1) without reasonable cause wholly or substantially suspending the carrying out of the contract works;
(2) failing to proceed regularly and diligently with the contract works;
(3) refusing or neglecting to comply with a written notice or instruction from the architect requiring the removal of any works, materials or goods not in accordance with the contract, where the contract works are materially affected by such refusal or neglect;
(4) failing to comply with the provisions of either clause 19.1.1. or 19.2.2 (which relate to assignation or sub-letting without written consent); or
(5) failing, pursuant to the contract conditions, to comply with the requirements of the Construction (Design and Management) Regulations 1994.

The employer's right to determine the contract arises on, or within ten days from, the expiry of the 14 day period. That right is exercised by serving a further notice upon the contractor. The determination takes effect on the date of receipt of that further notice. Notices are given by actual delivery, registered post or recorded delivery. Failure to comply with the provisions of the contract as to the giving of notice can be fatal, see for example *Muir Construction Ltd* v. *Hambly Ltd* (1990).

If the contractor remedies the specified default, or the employer elects not to serve the further notice required to determine, and the contractor repeats a specified default (whether he has previously repeated it or not), upon or within a reasonable time after such repetition the employer is entitled to serve the further notice required to determine. In these circumstances, the specified default need not continue for fourteen days, repetition alone is sufficient.

A notice served under clause 27 is not to be given unreasonably or vexatiously. For a consideration of the phrase 'unreasonably or vexatiously' see *J M Hill & Sons Ltd* v. *London Borough of Camden* (1980) and *John Jarvis* v. *Rockdale Housing Association* (1986).

Certain insolvency events will also entitle the employer to determine. The effect of insolvency is considered below, at section 8.8.

By virtue of clause 27.4 the employer is entitled to determine the employment of the contractor if the contractor has offered, given or agreed to give to any person any gift or consideration as an inducement.

The consequences of determination by the employer are specified in clauses 27.5 and 27.6. The contractual provisions as to determination by the employer are, by virtue of clause 27.6, without prejudice to any other rights and remedies the employer may possess. The other remedies open to the employer are considered below in Chapter 9.

### 8.4.3 Determination by contractor under JCT 98

The contractors' right to determine their employment is governed by clause 28 of JCT 98. The amendment effected to this clause by the Scottish Building Contract relates to the insolvency of the employer, and as such it applies only to the Private Editions of JCT 98. This provision is examined in more detail below at section 8.8.

The methods by which notices can be given are identical to those in relation to employer determination. Similarly, notices under this clause are not to be given unreasonably or vexatiously. As with employer determination, the clause sets out certain specified defaults. In addition, there are also what are termed specified suspension events, the occurrence of which can entitle contractors to determine their employment under the contract.

#### Specified defaults

Unlike determination by the employer, the specified defaults which entitle the contractor to determine their employment under the contract can arise both before and after practical completion. The specified defaults by the employer are set out in clause 28.2.1 and are:

(1) the employer not paying by the final date for payment the amount properly due to the contractor in respect of any certificate and/or any VAT due thereon;
(2) interfering with or obstructing the issue of any certificate;
(3) failing to comply with the provisions of clause 19.1.1 (that is assigning the contract without the written consent of the contractor); and
(4) failing to comply with the Construction (Design and Management) Regulations 1994.

#### Specified suspension events

The specified suspension events, which must arise prior to the date of practical completion, are set out in clause 28.2.2. This applies when the carrying out of the whole, or substantially the whole, of the uncompleted works is suspended for a continuous period of the length specified in the contract due to specified events, namely:

(1) the contractor not receiving in due time from the architect necessary instructions, drawings, details or levels;
(2) architect's instructions issued under clause 2.3 (discrepancies in or divergence between contract documents), clause 13.2 (instructions requiring a variation) or clause 23.2 (postponement of any work to be executed under the contract) unless caused by the negligence or default of the contractor, his servants or agents;
(3) delay in the execution of works not forming part of the contract which are to be carried out by the employer or persons engaged by the employer as referred to in clause 29; and
(4) failure to give in due time ingress to or egress from the site.

**Procedure and effect**

Once a notice has been given, the ensuing procedure is to all intents and purposes identical to that which operates in the case of employer determination, the only differences being the necessary modifications made to accommodate specified suspension events. Similarly, the provisions which deal with the repetition of a specified default or of a specified suspension event are virtually identical to those in employer determination. The consequences of determination by the contractor are specified by clause 28.4, which sets out a mechanism for determining the sum due to the contractor. The contractual provisions as to determination are without prejudice to any other rights and remedies which the contractor may possess.

Certain insolvency events entitle the contractor to determine. These will be considered below at section 8.8.

### 8.4.4 Determination by either party under JCT 98

Clause 28A of JCT 98 provides for certain circumstances which will entitle either the employer or the contractor to determine the employment of the contractor. Each of the specified circumstances must arise before the date of practical completion and must cause the carrying out of the whole or substantially the whole of the uncompleted works to be suspended for the relevant continuous period of time set out in the contract. The events provided for by clause 28A.1.1 are

(1) *force majeure*;
(2) loss or damage to the works occasioned by any one or more of the specified perils set out in clause 1.3;
(3) civil commotion, which has been defined as a stage between riot and war, see *Levy* v. *Assicurazioni Generali* (1940);
(4) architect's instructions issued as a result of the negligence or default of any local authority or statutory undertaker executing work solely in pursuance of its statutory obligations;

(5) hostilities involving the United Kingdom (whether or not war is declared); and
(6) terrorist activity.

Upon the occurrence of one or more of the aforementioned events, and once the period specified in the contract has expired, either party may give notice to the other to the effect that unless the suspension is terminated within seven days after receipt of that notice the employment of the contractor will be determined. The methods of giving notice are identical to those in the cases of employer and contractor determination and the notice cannot be given unreasonably or vexatiously. Unless the suspension is terminated within seven days after the date of receipt of the notice, the employment of the contractor is determined.

The contractor is not entitled to give notice where the loss or damage to the works occasioned by one or more of the specified perils is caused by some negligence or default on his part or on the part of his servants or agents, or any person employed or engaged upon or in connection with the works or their servants or agents, other than the employer or any person engaged by the contractor, or by any local authority or statutory undertaking executing work solely in pursuance of its statutory obligations. The consequences of determination under this clause are specified by clauses 28A.2 to 28A.7.

## 8.5 *Repudiation and rescission*

Notwithstanding the absence in a contract of detailed determination provisions, circumstances may arise whereby a party is freed from future performance. The concepts of repudiation and rescission are inextricably linked. Recission is considered in more detail below at Section 9.2. In certain circumstances a material breach of contract by one party may entitle the other party, the 'innocent party', to terminate the contract. Such a material breach of contract is referred to as a repudiation and gives the innocent party a choice. They can accept the repudiation and rescind the contract or, alternatively, they may elect to ignore the repudiation and continue with the performance of the contract. This option exists because, as stated by Lord Keith in the case of *Woodar Investment Development Ltd* v. *Wimpey Construction UK Ltd* (1980), the doctrine of repudiation exists for the benefit of the innocent party. However, whether in certain circumstances the option is restricted is considered below in Section 9.2.

If the repudiation is accepted the acceptance should be communicated to the party in breach. The method by which communication is made would appear to be immaterial, see *Monklands DC* v. *Ravenstone Securities* (1980).

The remedy that is open to an innocent party who elects to rescind, namely damages, is considered below at Section 9.4. Should the innocent party elect to ignore the repudiation they may well be barred from relying upon the material breach at a later date.

It is particularly difficult to generalise as to what conduct is, and what conduct is not, a repudiation. Not every material breach will constitute a repudiation, see *Blyth* v. *Scottish Liberal Club* (1982). Should one party refuse to perform their obligations under the contract that is likely to constitute a repudiation. Should an employer prevent a contractor from carrying out the contract works, for example by engaging another contractor to carry out all or part of those works, that too is likely to constitute a repudiation, see *Sweatfield Ltd* v. *Hathaway Roofing Ltd* (1997).

In Scotland, the precise effect of the acceptance of a repudiation and resultant rescission of the contract has been examined by the Inner House of the Court of Session in the case of *Lloyds Bank plc* v. *Bamberger* (1994) which provides a clear and succinct exposition of the position under Scots Law where a contract has been rescinded.

In *Lloyds Bank plc* Lord Ross stated that, following recission, both parties are freed from future performance of their primary obligations under the contract. Nevertheless parties continue to be bound by the primary obligations which are extant at the time of rescission. The contract does not come to an end. The innocent party is entitled to sue the party in default for damages for breach of contract. Ancillary clauses which the parties intended would survive rescission, such as arbitration clauses, may be enforced after rescission. Apart from such ancillary clauses, the contract may also contain clauses which affect damages due for breach of contract, such as a liquidated damages clause. The language of the contract may be such as to demonstrate that the parties intended such clauses to be enforceable after rescission.

Recission should be distinguished from contractual determination, considered above in section 8.4. In the latter, the employment of the contractor is determined, in certain instances upon the occurrence of events that are not the fault of either party. Contractual determination clauses seek to bring some degree of certainty to the circumstances in which the parties' contract comes to end prematurely.

## 8.6 Death and illness

The effect of the death or incapacity of a party to a building contract will primarily depend upon whether or not the contract involves an element of *delectus personae*. A contract that involves *delectus personae* means that one party to the contract entered into it in reliance upon certain qualities possessed by the other. Where such qualities are a necessary element of the contract the death or incapacity of the party who is bound to perform clearly prevents the contract being performed and, thus, brings it to an end. For examples of this see the cases of *Hoey* v. *McEwan & Auld and Others* (1867) and *Smith* v. *Riddell* (1886). The existence of *delectus personae* in a contract has a direct bearing upon whether or not that contract is capable of being assigned. This is considered below in Chapter 11.

The delegation of building work (through the use of sub-contractors) is an everyday occurrence and, therefore, in the absence of special circumstances or an express contractual provision to the contrary, *delectus personae* will not apply and the obligation to perform will pass to the personal representatives of the deceased party. It will be for the representatives to secure alternative contractors to carry out the works, or to complete them themselves!

As a building contract is ordinarily divisible (unlike a contract for a painting or a sculpture), it would appear that remuneration can be claimed by a deceased party's representatives for work partially carried out up to the date of death. The valuation of that work may be problematic, particularly if the contract does not have a mechanism for valuation, and may require equitable adjustment.

Illness and incapacity need to be treated in a like manner, although the position is, perhaps not surprisingly, not as clear cut as in the case of death. The effect of illness or incapacity is one of degree and will depend upon the whole circumstances, most notably the likely duration of the illness or incapacity in relation to the length of the contract. Even where illness is not sufficient to bring the contract to an end, it has been enough to entitle the employer to rescind the contract, see *Manson* v. *Downie* (1885), and to constitute a breach of contract, see *McEwan* v. *Malcolm* (1867).

Since the vast majority of contractors are now limited companies the problems occasioned by death and illness are unlikely to arise on a regular basis. However, it should be noted that *delectus personae* may arise in employer/contractor relationships that do not involve individuals as demonstrated by the case of *Scottish Homes* v. *Inverclyde DC* (1997). The issue may also arise in the case of architects or engineers appointed under a construction contract.

## 8.7 Illegality

While a detailed examination of the concept of illegality is beyond the scope of this work, it does merit some consideration.

In general terms, an illegal contract is one which the law will not enforce. However, there is a distinction between illegal contracts and those that are associated with an unenforceable transaction. Perhaps the best example that can be given to illustrate the latter is gambling. Whilst gambling is not illegal, the Scottish courts will not entertain actions to determine wagers. This is the case for reasons of public policy, not illegality. This long standing principle was recently confirmed in the case of *Ferguson* v. *Littlewoods Pools Ltd* (1997).

What precisely constitutes an illegal contract is open to question. A number of vague and differing concepts such as 'moral turpitude' and 'subversive of the interests of the State' have been used, see *Jamieson* v. *Watt's Trustee* (1950). The matter is far from clear, as is demonstrated by the decision in *Cuthbertson* v. *Lowes* (1870) in which it was held that a contract which contravened a statute was not necessarily illegal.

Whether or not either party questions the legality of the contract the court will have regard to it, see *F W Trevalion & Co* v. *J J Blanche & Co* (1919). Where a contract is held to be illegal the court will not interfere as between the rights of the parties to the contract. This is consistent with the general principle that the courts will not assist the party who is in breach of a statute, albeit that the corollary of this is that the other party to the contract is entitled to keep the advantage gained by them. This may be regarded as unfortunate where the parties were equally aware of the illegal nature of the transaction.

If the contract is not itself illegal, but has a connection with some other illegal transaction, the contract is said to be tainted with illegality. If one of the parties was unaware of the illegality they will be entitled to enforce their rights under the contract. However, should they fail to resile from the contract, after becoming aware of the illegality, they may be held to have acquiesced. As a consequence, they may not be entitled to enforce their rights under the contract.

Where only part of a contract is illegal that part may be capable of being severed from the remainder of the contract. In a building contract which contains the power to instruct variations it may well be possible to instruct a variation to remove the 'offending' part of the contract. The question of severance is a complex one upon which there is little Scottish authority, although the English authorities on this subject are likely to be regarded as highly persuasive. Whether or not an illegal provision is capable of being severed will depend upon the nature of the illegality.

The contract may be valid, but the works executed under it illegal. For example, in *Townsend (Builders) Ltd* v. *Cinema News and Property Management Ltd* (1959) the works as built, but not as specified, contravened a byelaw. In the rather special circumstances of that case the contractor was held to be entitled to recover payment, although it would appear that but for those special circumstances the contractor would not have succeeded.

If contractors carry out work in the absence of necessary consents they take the risk that the work is illegal and they may be unable to recover their charges for that work, see *Designers and Decorators (Scotland) Ltd* v. *Ellis* (1957). The contract will often expressly oblige the contractor to give statutory notices and comply with statutory requirements, and may also provide for the circumstances in which there is a change in the statutory requirements. For example, clause 6 of JCT 98 makes such provision.

## 8.8 *Insolvency*

### 8.8.1 General

Insolvency, in itself, does not affect a contract, but has potentially far reaching implications that merit some examination within the confines of this work.

The insolvent party may be unable to implement their obligations under the contract, which would entitle the other party to withhold performance of their obligations under the contract, see *Arnott and Others* v. *Forbes* (1881). In a case of personal insolvency (known as bankruptcy) the party contracting with the insolvent debtor can compel the debtor's representative (known as a permanent trustee) to make his position clear in relation to the contract.

Section 42 of the Bankruptcy (Scotland) Act 1985 provides that the permanent trustee is deemed to have refused to adopt the contract unless he responds within 28 days from the receipt by him of a request in writing from any party to a contract entered into by the debtor, or within such longer period of that receipt as the court on application by the permanent trustee may allow.

### 8.8.2 Effect of insolvency under JCT 98

With the exception of the appointment of a provisional liquidator, the insolvency of the contractor forthwith automatically determines his employment under a JCT 98 contract. The Scottish Building Contract deletes clause 27 of JCT 98 entirely and substitutes a new clause for use in Scotland. Insolvency events are dealt with by clause 27.3.

In the event of a provisional liquidator being appointed to control the affairs of the contractor, the employer has the option to determine the contractor's employment. The employer is entitled to determine by giving the provisional liquidator seven days written notice of such determination.

Clause 27.3.2 deals with automatic determination on the contractor's insolvency. Determination occurs automatically in the event of contractors:

- becoming bankrupt or making a composition or arrangement with their creditors or
- having a proposal for a voluntary arrangement for a composition of debts or scheme of arrangement approved in accordance with the Insolvency Act 1986 or
- having an application made under the Insolvency Act 1986 to the court for the appointment of an administrator or,
- being individuals or a Scottish partnership, having their estate sequestrated or becoming apparently insolvent (as defined by the Bankruptcy (Scotland) Act 1985) or
- entering into a trust deed for their creditors or
- having a winding up order made
- or (except for the purposes of reconstruction) having a resolution for voluntary winding up passed
- or having a receiver or manager of their business or undertaking duly appointed or
- having an administrative receiver (as defined in the Insolvency Act 1986) appointed.

Determination is also automatic if possession is taken by or on behalf of the holder of any debenture secured by a floating charge.

In the event of the employment of the contractor being determined by the employer, or automatically determined due to the insolvency of the contractor, that employment may be reinstated and continued if the employer and the contractor, their trustee in bankruptcy, provisional liquidator,

liquidator, receiver, manager or administrative receiver (as the case may be) shall so agree. Where the employment of the contractor under the contract has been determined by virtue of the contractor's insolvency, and it has not been reinstated and continued, the rights and duties of both the employer and the contractor are set out by clause 27.5.

The SBCC has produced a practice guide for employers and their professional advisers upon the event of the insolvency of a main contractor while carrying out work under a Scottish Building Contract. This work is commended to readers for a fuller examination of the subject.

Determination by the contractor in the event of the insolvency of the employer will only affect the Private Editions of the JCT 98. The Scottish Building Contract deletes clause 28.3.1 of JCT 98 and replaces it with a new clause for use in Scotland, which defines the circumstances in which the clause applies. With minor inconsequential differences in wording, the new clause 28.3.1 is identical to clause 27.3.2, which deals with the insolvency of the contractor.

## 8.9 Prescription

### 8.9.1 General

Prescription is the establishment or definition of a right or the extinction of an obligation through the lapse of time. The former is termed positive prescription, the latter negative prescription. Positive prescription applies to title to interests in land, servitudes and public rights of way. It has no relevance to building contracts and will not be considered further in this book.

Prescription falls to be contrasted with limitation. Limitation does not affect the subsistence of rights and obligations. It is a doctrine that denies certain rights of action after the passage of a certain lapse of time, see *Macdonald* v. *North of Scotland Bank* (1942). Limitation periods may be statutory or conventional. Conventional limitation is where the parties set out in their contract that a particular obligation will be extinguished by the lapse of a stipulated time period without a claim being made.

Such a provision can appear in construction contracts. For example, clause 15(6) of the Federation of Civil Engineering Contractors Form of Sub-Contract (universally known as the 'Blue Form') provides that the contractor has no liability to the sub-contractor for any matter or thing arising out of or in connection with the sub-contract or the execution of the sub-contract works unless the sub-contractor has made a written claim in respect of that matter or thing to the contractor before the engineer issues the maintenance certificate in respect of the main contract works.

The law in relation to both prescription and limitation is to be found in the Prescription and Limitation (Scotland) Act 1973 (which we will refer to in this chapter as 'the 1973 Act'). Whilst a detailed examination of this subject is beyond the scope of this work, we shall consider those aspects of it which are most pertinent to building contracts, namely short negative prescription and long negative prescription.

Under the 1973 Act the party under the relevant obligation is known as 'the debtor' and the party to whom the obligation is owed is known as 'the creditor'. In relation to the short negative and the long negative prescriptive periods, the general rule is that the burden of proof in establishing whether or not an obligation has prescribed rests with the party alleging the affirmative. For example, if the assertion is that the obligation had subsisted for the prescriptive period it would be for the party so affirming to prove, see *Strathclyde Regional Council* v. *W A Fairhurst & Partners* (1997).

### 8.9.2 Short negative prescription

The short negative prescriptive period of five years is the one most familiar to those in the construction industry. Section 6 (1) of the 1973 Act provides that if, after the 'appropriate date', an obligation which is set in out in Schedule 1 to the 1973 Act has subsisted for five years (a) without any relevant claim having been made in relation to it, and (b) without the subsistence of the obligation having been relevantly acknowledged, then as from the expiration of the five year period the obligation in question is extinguished.

A number of technical expressions are used in s.6(1). As we will see below, many of these are equally relevant to long negative prescription. We shall examine each of these expressions in turn.

### 8.9.3 The appropriate date

The short negative prescriptive period commences upon what is termed the 'appropriate date'. This date varies from obligation to obligation. Schedule 2 to the 1973 Act sets out various obligations and the appropriate date relative to each of them. None of the Schedule 2 obligations are particularly relevant to building contracts. With the exception of obligations of the kind specified in Schedule 2, the appropriate date in relation to an obligation is the date upon which that obligation became enforceable.

Section 11 of the 1973 Act defines when certain types of obligation become enforceable. For example, an obligation to make reparation for loss, injury or damage caused by an act, neglect or default is regarded as having become enforceable on the date when the loss, injury or damage occurred or was discovered.

There must be an act, neglect or default and resultant loss, injury or damage. The obligation to make reparation does not arise, and thus does not become enforceable, until the loss, injury or damage occurs, see *Watson* v. *Fram Reinforced Concrete Co (Scotland) Ltd* (1960), *Dunlop* v. *McGowans* (1979) and *Strathclyde Regional Council* v. *W A Fairhurst & Partners* (1997).

The loss, injury or damage must arise from the act neglect or default. For example, in *Sinclair* v. *MacDougall Estates Ltd* (1994) it was held that the defenders' act, neglect or default founded upon by the pursuers was not a breach of the general duty to construct in accordance with the contract, but was constituted by certain specified failures on the defenders' part to design and construct the building in a workmanlike manner in terms of the contract. The minor breaches of the contract which had caused damage discovered at earlier stages (in 1972 or 1977) were not sufficient to constitute *injuria* in relation to major and different failures to design and construct the building properly which had resulted in the damage discovered in 1988. The loss, injury or damage sustained in 1988 did not arise from the act, neglect or default discovered in 1972 or 1977 and the case was held not to be time barred.

A number of the relevant cases on this subject arise from building contracts and these usefully illustrate the position. In *George Porteous (Arts) Ltd* v. *Dollar Rae Ltd* (1979) contractors were refused planning permission and the work executed by them had to be demolished. In that case it was held that the prescriptive period ran from the date of service of the enforcement notice, that being the date upon which the pursuers suffered loss. In *Scott Lithgow Ltd* v. *Secretary of State for Defence* (1989) the prescriptive period was held to have commenced as from the date when the materials in question were found to have been defective.

Section 11(2) of the 1973 Act provides that where, as a result of a continuing act, neglect or default, loss has occurred prior to the act, neglect or default ceasing, the loss is deemed to have occurred on the date when the act ceased.

Where the creditor is not aware, and could not with reasonable diligence have been aware that loss, injury or damage has occurred, the prescriptive period does not commence until the date on which the innocent party first became, or could with reasonable diligence have become so aware, see s.11 (3). This provision has particular relevance in the case of latent defects; the five year period will commence from the date of 'discoverability' of the defect, subject to the 'long-stop' of the 20 year long negative prescriptive period, which is considered below at section 8.9.7.

### 8.9.4 Schedule 1 obligations

The type of obligations that are affected by the short negative prescriptive period are defined in Schedule 1 to the 1973 Act. Unlike long negative prescription, short negative prescription applies only to a limited number of obligations. Of these, certain are particularly relevant to building contracts. These are set out in paragraph 1 of Schedule 1 and are any obligation

- based on unjustified enrichment (including restitution, repetition or recompense),
- arising from liability to make reparation and
- arising from, or by reason of any breach of, a contract or promise, not being an obligation falling within any other provision of paragraph 1.

An obligation arising under a contract will include an obligation to refer disputes under an engineering contract to the contract engineer, see *Douglas Milne Ltd* v. *Borders Regional Council* (1990). The same will apply in the case of an arbitration clause. A performance bond has been held to be a cautionary obligation subject to the short negative prescriptive period, see *City of Glasgow DC* v. *Excess Insurance Co Ltd* (1986). The appropriate date in such a case is the date of issue of an architect's certificate ascertaining the extent of the damages due for default, see *McPhail* v. *Cunninghame DC* (1985) and *City of Glasgow DC* v. *Excess Insurance Co Ltd (No.2)* (1990).

### 8.9.5 Relevant claims

An obligation affected by short negative prescription will be extinguished if it has subsisted for a continuous period of five years without either of two events occurring, namely, the making of a relevant claim or a relevant acknowledgement.

If a relevant claim is made the prescriptive period is said to have been interrupted and a new five year period commences as from the date of interruption, see s.9.

A 'relevant claim' is one made by or on behalf of the creditor in an obligation for implement or part implement of the obligation in 'appropriate proceedings' or in certain insolvency related circumstances. An examination of the latter is beyond the scope of this work. 'Appropriate proceedings' means

- court proceedings in Scotland,
- an arbitration in Scotland or
- an arbitration outside Scotland in which an award would be enforceable in Scotland.

The date of the relevant claim is the date of service of court proceedings, except in relation to Court of Session proceedings which do not subsequently call. If Court of Session proceedings do not call a relevant claim is not made. In the case of an arbitration, the date of the relevant claim is the date when the claim is made in the arbitration or the preliminary notice is served, whichever is the earlier. If no preliminary notice is served the relevant claim in an arbitration will be made on the date when the claim is actually made. To be a relevant claim the preliminary notice must state the nature of the claim, see *Douglas Milne Ltd* v. *Borders Regional Council* (1990).

### 8.9.6 Relevant acknowledgements

Section 10 of the 1973 Act defines a relevant acknowledgement. The subsistence of an obligation is regarded as having been relevantly acknowledged if, and only if, either of two defined conditions is satisfied. Firstly, there must have been such performance by or on behalf of the debtor towards implement of the obligation as clearly indicates that the obligation still subsists. Secondly, and alternatively, there has to have been made by or on behalf of the debtor to the creditor or his agent an unequivocal written admission clearly acknowledging that the obligation still subsists. As with a relevant claim, if a relevant acknowledgement is made the prescriptive period is interrupted and a new five year period commences as from the date of interruption.

### 8.9.7 Long negative prescription

In Scotland, the long negative prescriptive period is 20 years. Section 7 (1) of the 1973 Act provides that if, after the date when an obligation became enforceable, the obligation has subsisted for a continuous period of 20 years without either a relevant claim or a relevant acknowledgement, then as from the expiration of the 20 year period the obligation is extinguished. Long negative prescription does not apply to obligations arising under s.22A of the 1973 Act (liability under the Consumer Protection Act 1987 for a defect in a product) or specified in Schedule 3 to the 1973 Act (imprescriptible rights and obligations), or to obligations under Schedule 1 of the 1973 Act to which the short negative prescriptive period applies.

Other than the length of the period the main difference between the short negative and long negative prescriptive periods is the point in time at which they commence. As we have seen the former commences as from the appropriate date. The latter commences as from the date upon which the obligation in question became enforceable.

The practical consequence of this in the context of building contracts is significant. The concept of 'discoverability' of a latent defect (which applies to the five year period) does not apply to the 20 year period, and thus an obligation arising from a latent defect will, in the absence of a valid interruption of the prescriptive period, prescribe 20 years after the date upon which the obligation became enforceable which is, broadly, when there has been both an act, neglect or default and loss, injury or damage arising therefrom. Thus a latent defect that is discovered 19 years after the obligation in question became enforceable will prescribe 20 years after the date upon which the obligation became enforceable i.e. in these circumstances only one year after the discovery. A defect that is discovered less than 15 years after it became enforceable will prescribe five years after it is discovered.

It should be noted that, in long negative prescription, there is no equivalent provision to s.11 (3) of the 1973 Act. Accordingly, even if the creditor was not aware, and could not with reasonable diligence have been aware, that loss, injury or damage had been caused, the long negative prescriptive period continues to run. Discoverability is not an issue in long negative prescription.

As with short negative prescription, the long negative prescriptive period can be interrupted by the making of a relevant claim or by a relevant acknowledgement. In this regard, the provisions of sections 8.9.5 and 8.9.6. apply equally to long negative prescription.

# Chapter 9
# Remedies

## 9.1 *Introduction*

Disputes arise under building contracts as with any other type of contract. While the resolution of such disputes is considered in Chapter 14, in this chapter we will consider certain of the remedies that are open to parties where a dispute arises.

Whilst certain of the remedies are, perhaps, peculiar to building contracts, the ordinary remedies that are open to the parties to any form of commercial contract are available. The remedies that are most commonly associated with building contracts are to be found within the provisions of the standard form contracts, such as JCT 98. Certain of these remedies, such as liquidated and ascertained damages and extensions of time (Chapter 5), and determination (Chapter 8) are considered elsewhere within this book. However, certain others are considered here. Separately, we will consider the general common law remedies open to parties, certain of which are quite independent of those arising under the terms of a specific contract.

Ordinarily, the general common law remedies and the remedies provided for in a specific contract will exist at the same time, see *Gilbert Ash (Northern) Ltd* v. *Modern Engineering (Bristol) Ltd* (1974). A party's common law rights can only be taken away by clear, unequivocal words, see *Redpath Dorman Long Ltd* v. *Cummins Engine Co Ltd* (1981). The extent to which that is achieved will depend upon the terms of the contract in question. See, for instance the Scottish case of *Eurocopy Rentals Ltd* v. *McCann Fordyce* (1994) in which it was held that the contractual termination provision was the exclusive method of termination.

There are both advantages and disadvantages associated with each of the types of remedy. For example, a liquidated damages provision of the nature contained in clause 24 of JCT 98 is of advantage to the employer in that it is not required to prove the actual loss it has sustained as a result of the contractor failing to complete the works on time. The downside of clause 24 is that employers must adhere strictly to the provisions of the clause to entitle them to deduct liquidated damages.

## 9.2 *Rescission*

The concepts of repudiation and rescission were considered in Section 8.5. In certain circumstances a material breach of contract by one party may be such as to entitle the other party (the 'innocent party') to terminate the contract. If the breach constitutes a repudiation of the contract, the innocent party has a choice. They can either accept the repudiation and rescind the contract or, alternatively, they may elect to ignore the repudiation and continue with the performance of the contract. The extent to which there is a right to continue with performance may, however, be limited.

The remedy open to the innocent party is to rescind the contract. Where a contract has been rescinded, both parties are freed from future performance of their primary obligations thereunder. Parties continue to be bound by the primary obligations that were extant at the time of rescission. The contract does not come to an end. The innocent party is entitled to sue the party in default for damages for breach of contract. Ancillary clauses which the parties intended would survive rescission, such as arbitration clauses, may be enforced after rescission. Apart from such ancillary clauses, the contract may also contain clauses which affect the amount of damages due for breach of contract, such as a liquidated damages clause. The language of the contract may be such as to demonstrate that the parties intended such clauses to be enforceable after rescission, see *Lloyds Bank plc* v. *Bamberger* (1994).

It is well established law that the innocent party's remedies for the other party's breach of contract are limited to those provided for in the contract, and for those breaches committed while the contract subsisted. However, where the contract has been rescinded and the mutual obligations thereunder have ceased to exist, the court has the power to award compensation on the basis of *quantum meruit*, see *Morrison-Knudsen Co Inc* v. *British Columbia Hydro and Power Authority* (1991), approved by the Inner House of the Court of Session in *ERDC Construction Ltd* v. *H M Love & Co.* (1995). The subject of payment *quantum meruit* is considered in Section 7.4.

The ordinary position in contract is that the innocent party is entitled to ignore the repudiation and continue with the performance of the contract. This is supported by the decision of the House of Lords in *White & Carter (Councils) Ltd* v. *McGregor* (1962) and, in the case of building contracts, by the decision of the Inner House of the Court of Session in *ERDC Construction Ltd.* The courts have, however, recognised that there may be limitations upon the right of the innocent party to insist upon performance. In *White & Carter (Councils) Ltd,* Lord Reid made it clear that:

> '... had it been necessary for the defender to do or accept anything before the contract could be completed by the pursuers, the pursuers could not and the court would not have compelled the defender to act, the contract would not have been completed, and the pursuers' only remedy would have been damages.'

It must be noted that in *White & Carter (Councils) Ltd* the pursuers did not require any co-operation, either active or passive, on the part of the defender. Building contracts patently cannot be performed without co-operation between employer and contractor. The courts in England have made it clear that they will not enforce an agreement for two people to live peaceably under the same roof, see *Thompson* v. *Park* (1944). The rationale behind this is identical to that identified by Lord Reid in *White & Carter (Councils) Ltd.* A multitude of practical problems would arise if the courts compelled performance.

In these circumstances, the authors respectfully suggest that the *obiter* comments of Lord Reid in *White & Carter (Councils) Ltd* should apply in relation to building contracts and the innocent party's right to insist upon performance is limited. Support for this proposition is to be found in the case of *London Borough of Hounslow* v. *Twickenham Garden Developments Ltd* (1970). In essence, the innocent party may be forced to accept a repudiation and rescind the contract, see *Decro-Wall International SA* v. *Practitioners in Marketing Ltd* (1971). The nature of building contracts is such that the option to ignore an employer's repudiation may not be one that is open to a contractor.

## 9.3 *Specific implement*

When parties enter into a contract, they each undertake certain obligations. In Scotland, it is presumed that contractual obligations will be enforced by the courts, unless there are considerations which make implement impossible or unjust, see *Stewart* v. *Kennedy* (1890) and *Beardmore* v. *Barry* (1928). The remedy open to the innocent party to compel performance of contractual obligations by the party in breach is known as specific implement.

As with recission, the innocent party, in most circumstances, has a choice. Either they can insist upon their entitlement under the contract, or they can seek damages for the breach, see *Holman & Co* v. *Union Electric Co* (1913). Damages are considered below in section 9.4. It will, however, always be at the discretion of the court as to whether the remedy of specific implement or that of damages is the appropriate one, see *Graham* v. *Magistrates and Police Commissioners of Kirkcaldy* (1881). Specific implement is not an appropriate remedy in every case. It is not available in respect of the enforcement of a party's monetary obligations under a contract, see *White & Carter (Councils) Limited* v. *McGregor* (1962). It is also an inappropriate remedy where performance would require the party in breach to become a partner in a commercial undertaking, see *Pert* v. *Bruce* (1937). It is an inappropriate remedy where performance is impossible, see *McArthur* v. *Lawson* (1877). There are a number of other instances in which specific implement has been held to be inappropriate. A more detailed examination of these is beyond the scope of this book.

The issue of specific implement in the context of building contracts is a difficult one. It raises similar issues to those that arise in relation to rescission (see section 9.2). In *London Borough of Hounslow* v. *Twickenham Garden Developments Ltd* (1970) it was argued on behalf of the borough that the contract (which incorporated the RIBA conditions) was not specifically enforceable. Whilst a decision on this point was not necessary to resolve the case, Megarry J (as he then was) stated that he could not see why the contract should not be held to be specifically enforceable. In contrast to rescission, whether or not specific implement is an appropriate remedy in relation to a particular contractual obligation will depend upon whether the party in breach is required to do, allow or accept something. Such co-operation may be essential in relation to certain obligations, but unnecessary in respect of others.

In the context of building contracts, the authors respectfully submit that where co-operation by the party in breach is required, specific implement is not an appropriate remedy. Such an approach is consistent with the comments of Lord Reid, and indeed the dissenting opinion of Lord Morton of Henryton, in *White & Carter (Councils) Ltd.* The innocent party must have an interest to insist upon specific implement of the obligation. If the court is not satisfied that they have such an interest their claim will be for damages, see *Clea Shipping Corp* v. *Bulk Oil International Ltd* (1984). The reality is that, in the majority of cases, the innocent party in a building contract may have no legitimate interest in performing the contract, rather than claiming damages. In such cases, it has been suggested that the innocent party can, in one sense, be said to be forced to claim damages. To insist upon any other remedy would be of little value, see *Decro-Wall International SA* v. *Practitioners in Marketing Ltd* (1971). As stated by Sachs LJ in *Decro-Wall International SA*, 'in such cases it is the range of remedies that is limited, not the right to elect'.

In Scotland, the remedy of specific implement is available both in the Court of Session and in the Sheriff Court. By virtue of s.47(2) of the Court of Session Act 1988, interim orders for specific implement can competently be granted. Such a remedy is not available in the Sheriff Court. Where a party to an adjudication under the Scheme for Construction Contracts (Scotland) Regulations 1998 declines to comply with an adjudicator's decision, or a request to register the decision for execution in the Books of Council and Session, specific implement may be used to compel performance. Where the adjudication has proceeded under the Scheme for Construction Contracts (Scotland) Regulations 1998, and there is such a failure to comply by one of the parties to it, it is also open to the innocent party to seek an order for specific performance under s.45 (b) of the Court of Session Act 1988. This entitles the court to order the specific performance of any statutory duty, under such conditions and penalties (including fine and imprisonment, where consistent with the enactment concerned) in event of the order not being implemented, as the court sees fit.

## 9.4 *Damages for breach of contract*

### 9.4.1 General

Perhaps the most commonly used legal remedy is that of damages. Damages are expressed in monetary terms. The purpose of damages for breach of contract is to place the innocent party in the position they would have been in had the breach not occurred. Damages may also be recoverable where no contract exists but one party owes the other a duty of care and is in breach of that duty. In those circumstances damages will be recoverable under the law of delict rather than for breach of contract. Delictual claims are considered below in section 9.10.

It must be appreciated that the law of damages is a vast and complex subject and a detailed examination is beyond the scope of this work.

It is open to the parties to a contract to decide in advance what damages, if any, will be payable in the event of a breach by one, or any of them. In essence, it is open to parties to exclude or limit liability. This is a common occurrence in building contracts, for example clause 24 of JCT 98, which provides for the payment of liquidated and ascertained damages for non-completion (see Section 5.9). It should also be borne in mind that claims for damages are subject to the law of prescription (see Section 8.9).

### 9.4.2 Causation, foreseeability and remoteness

The loss which the pursuer is entitled to recover is that which has been caused by the defender's breach, see *Bourhill* v. *Young* (1942). The pursuer must establish not only that there has been a breach of contract but also that there is a causal connection between the breach and the losses sought to be recovered, and that such losses are not too remote.

Only losses that were foreseeable as being the likely consequences of a breach at the time the contract was entered into are recoverable. If the losses were not foreseeable at that time, they are too remote and cannot be recovered.

These issues were considered in the case of *Hadley* v. *Baxendale* (1854). This case laid down the rules that apply in assessing the measure of damages in a breach of contract case. In delivering the judgment of the court, Alderson B stated:

> 'Where two parties have made a contract which one of them has broken, the damages which the other party ought to receive in respect of such breach of contract should be such as may fairly and reasonably be considered either arising naturally, i.e., according to the usual course of things, from such breach of contract itself, or such as may reasonably be supposed to have been in the contemplation of both parties, at the time they made the contract, as the probable result of the breach of it.'

That part of the judgment of the court has been widely repeated and relied upon since 1854. In the context of building contracts, however, the following part of the judgement is also significant. Alderson B went on to state:

> '[I]f the special circumstances under which the contract was actually made were communicated by the [pursuers] to the [defenders], and thus known to the both parties, the damages resulting from the breach of such a contract, which they would reasonably contemplate, would be the amount

of injury which would ordinarily follow from a breach of contract under these special circumstances so known and communicated. But on the other hand, if these special circumstances were wholly unknown to the party breaking the contract, he, at the most, could only be supposed to have had in his contemplation the amount of injury which would arise generally ... from such a breach of contract.'

An example of such special circumstances is to be found in *Balfour Beatty Construction (Scotland) Ltd* v. *Scottish Power plc* (1994). In that case the pursuers were engaged in the building of a roadway and associated structures, including an aqueduct. They contracted with the defenders' predecessors for the supply of electricity to operate a concrete batching plant. The construction of the aqueduct required a continuous pour operation. In the course of the construction of the aqueduct, the batching plant stopped working. It was established that the electricity supply had been interrupted and that the interruption was a breach of contract by the defenders' predecessors. The construction company claimed the cost of demolishing and re-building a substantial part of their works, this having been rendered necessary by the interruption of the electricity supply and the consequent interruption of the required continuous pour. It was established that the defenders' predecessors had not known of the need for a continuous pour. The Lord Ordinary concluded that the need to re-build part of the works as a result of the interruption of the continuous pour had not been within the defenders' reasonable contemplation and the action failed. Ultimately, the House of Lords upheld the decision of the Lord Ordinary. Had the special circumstances, namely, the need for a continuous pour, been known to the defenders, it is likely that the pursuers would have succeeded.

### 9.4.3 Damages recoverable and mitigation

The law of Scotland is clear in respect of the method of assessment of damages, assuming the necessary prerequisites considered above at section 9.4.2 have been met. As was stated by Lord Pearson in *The Govan Rope & Sail Co Ltd* v. *Andrew Weir & Co* (1897):

> '[I]t appears to me that the criterion of damage now adopted by the pursuers is in accordance with the principle which governs the whole law on the subject, namely, that the party observing the contract is to be put as nearly as possible in the same position as he would have been if the contract had been performed.'

As damages is a monetary remedy, the party suffering the loss can only be put in the position it would have been in, but for the breach, insofar as a payment of money to them allows.

Building contracts will often confer on the contractor a right to an increase in the contract sum for direct loss and/or expense incurred by the contractor on the occurrence of certain events, e.g. failure or delay in issuing instructions or information, which otherwise would be treated as damages for breach of contract. Although the nature of the claim in such cases is conceptually different insofar as it is a claim for payment under the contract rather than for damages, the calculation of the 'direct loss and/or expense' is likely to be little different from a claim for damages for breach of contract, see JCT clause 26 considered above in Section 7.3.

If the contract is rescinded by the contractor due to a material breach by the employer, the contractor is likely to be entitled to recover by way of damages the profit he would have made had the contract been completed in the ordinary course. If the works had been partially carried out, the contractor retains the right to payment for the value of such works. The rescission of a contract in consequence of a repudiation does not affect accrued rights to payment under the contract, unless the contract provides that it was to do so, see *Hyundai Heavy Industries Co Ltd* v. *Papadopoulos* (1980).

In the event of a breach of contract by the contractor, again the contract may provide the remedy in certain circumstances, see for example the liquidated and ascertained damages provisions in respect of non-completion under clause 24 of JCT 98 which are considered above at Section 5.9. If the contractor does not complete the contract works the employer's loss will be the additional cost of completing the works, if any. If the works are completed at no additional cost there will be no loss. It should, however, be noted that there is authority in Scotland to the effect that where a breach of contract is established the pursuer is entitled to nominal damages, even if no loss can be demonstrated. This comes from the opinion of the Lord President in *Webster & Co* v. *The Cramond Iron Co* (1875) in which he stated that:

> '[Where the] contract and the breach of it are established ... that leads of necessity to an award of damages. It is impossible to say that a contract can be broken even in respect of time without the party being entitled to claim damages - at the lowest, nominal damages.'

Another common breach of contract by the contractor is the existence of defects in the works executed or that the works executed do not conform to the requirements of the parties' contract. Assuming the contract contains no specific mechanism under which the contractor is obliged to remedy defects, such as clause 17 of JCT 98, and the contractor will not do so voluntarily, the measure of the employers' loss can, ordinarily, be assessed in one of two ways. The first is the cost of the necessary repairs. The second is the difference in value between the building in the condition contracted for and the building in its actual condition, i.e. with the defective work, see *GUS Property Management Ltd* v. *Littlewoods Mail Order Stores Ltd* (1982).

A pursuer can proceed on the basis of either measure. These are not the only available measures of loss and a court is not confined to making an award based on one of these measures, see *Ruxley Electronics and Construction Ltd* v. *Forsyth* (1995). It may be prudent, where possible, to proceed on the basis of both measures as alternatives. The proper measure of damages may be determined by checking one measure against the other, see *Prudential Assurance Co Ltd* v. *James Grant & Co (West) Ltd* (1982).

The law in relation to this matter has recently been clarified (at least in England) in the case of *Ruxley Electronics and Construction Ltd*. The House of Lords held that in assessing damages for breach of contract for defective building works, should the court decide that the cost of reinstatement would be out of all proportion to the benefit to be obtained to the innocent party by reinstatement, the innocent party's claim would be restricted to the difference in value between the building in the condition contracted for and the building in its actual condition. Whether or not the innocent party actually intends to reinstate will be relevant in determining if it is reasonable to insist upon reinstatement.

It should be borne in mind that certain contracts have detailed mechanisms for assessing the sum due by one party to the other on determination, see for example clauses 27, 28 and 28A of JCT 98, as amended by the Scottish Building Contract. In part, at least, this deals with damages arising out of the determination. Determination is considered above in Section 8.4.

Finally, it must always be borne in mind that there is incumbent upon a pursuer the duty of taking all reasonable steps to mitigate the loss consequent on the breach. The pursuer is not entitled to claim for losses caused by their failure to take such steps, see *British Westinghouse Electrical & Manufacturing Co Ltd* v. *Underground Electric Railways Co of London Ltd* (1912).

## 9.5 Finance charges

It has long been judicially recognised that, in the ordinary course of things, when contractors require capital to finance a contract they either borrow the capital and pay for the privilege, or use their own capital and, as a consequence, lose the interest which they would otherwise have earned, see for example *F G Minter Ltd* v. *Welsh Health Technical Services Organisation* (1980). Similarly, it has been judicially recognised that, in the construction industry, delay in payment to contractors might naturally result in them being short of working capital, thus causing them to incur finance charges, see for example *Ogilvie Builders Ltd* v. *City of Glasgow DC* (1995). Whether or not such finance charges are recoverable by contractors has been the subject of considerable judicial discussion over the years but it is now well settled, both in Scotland and England, that finance charges are recoverable in certain defined circumstances.

Firstly, finance charges are recoverable as direct loss and/or expense under clause 26 of JCT 98, but only if the requirements of that clause are satisfied, see *Ogilvie Builders Ltd*, following *F G Minter Ltd* and *Rees & Kirby Ltd* v. *Swansea City Council* (1985). Clause 26 of JCT 98 is considered in Section 7.3 above.

Secondly, finance charges are recoverable as damages in a case based on breach of contract. The words 'direct loss and/or expense' are to be given the same meaning in a case of breach of contract as would be given in a case for payment under contract, see *Ogilvie Builders Ltd*. Recovery by way of a claim based upon breach of contract has (at least until the decision in *Ogilvie Builders Ltd*) proved more problematic in Scotland, with claims being unsuccessfully advanced in cases such as *Chanthall Investments Ltd* v. *F G Minter Ltd* (1975). It was stressed in that case, however, that, in each case where this issue arises, it is a question of fact and the particular circumstances as to whether or not the loss in question was within the contemplation of the parties. This approach was approved by the Inner House of the Court of Session in *Margrie Holdings Ltd* v. *City of Edinburgh DC* (1994). This approaches recovery by way of the second branch of what is known as the rule in *Hadley* v. *Baxendale* (1854), namely, losses such as may reasonably have been supposed to have been in the contemplation of both parties, at the time the contract was entered into, as the probable result of a breach of it. This falls to be contrasted with the first branch of the rule, namely, that where two parties have entered into a contract and there has been a breach of contract by one of the parties, the damages to which the innocent party is entitled should be such as may fairly and reasonably be considered as arising naturally from the breach, see section 9.4.2 above.

In *Ogilvie Builders Ltd* Lord Abernethy stated that he did not read any of the Scottish cases cited to him as indicating any general proposition that claims for finance charges, if recoverable at all, could only be recoverable under the second branch of the rule in *Hadley* v. *Baxendale*, holding that a claim for finance charges under the first branch was relevant, as a matter of law. In Scotland, claims advanced under the second branch of the rule have been held to be relevant as a matter of law, see *Caledonian Property Group Ltd* v. *Queensferry Property Group Ltd* (1992). What *Ogilvie* recognised was the commercial reality that extra finance charges could arise 'naturally' from a breach of contract in the construction industry.

## 9.6 Interest

### 9.6.1 Common law

Late payment of sums admittedly due is commonplace in the construction industry. One way in which this can be addressed is by way of interest. The general rule in Scotland is that, unless a contract provides otherwise, interest will only be awarded from a date prior to the serving of a writ if the money has been wrongfully withheld. That has been the position in Scotland for some considerable time and was enunciated by Lord Atkin in *Kolbin & Sons* v. *Kinnear & Co* (1931). His Lordship stated that:

> '[I]t seems to be established that, by Scots Law, a pursuer may recover interest by way of damages where he is deprived of an interest-bearing security or a profit-producing chattel, but otherwise, speaking generally, he will only recover interest, apart from contract, by virtue of a principal

sum having been wrongfully withheld and not paid on the day where it ought to have been paid.'

The observations of Lord Atkin were accepted in *F W Green & Co* v. *Brown & Gracie Ltd* (1960) as setting out the broad principle. Whilst a number of cases have dealt with the issue of interest, none of them has ever precisely said what is meant by 'wrongfully withheld'. Various views have been expressed as to its meaning, including failure to pay following the issuing of a certificate, 'negligent' under-certification and client interference in the certification process. The law of Scotland in relation to the entitlement to interest has recently been commented upon by the Inner House of the Court of Session in *Elliott* v. *Combustion Engineering Ltd* (1998).

Whilst a decision of the Inner House of the Court of Session on this topic is welcome, it must be said that the opinion of the court in *Elliott* does not, in reality, answer many of the questions that have been posed in this field over the years. The difficulty with the decision is that, perhaps understandably, 'wrongful withholding' is not directly defined and the court's conclusion was that modern authority indicated that, in general, interest would run on contractual debts from judicial demand (that is service of a writ), and that while there might be qualifications or exceptions to the general rule, the circumstances of *Elliott* did not fall within any such qualification or exception.

Unfortunately, the extent of the qualifications or exceptions to the general rule is not fully set out in *Elliott*. It must, however, be recognised that this was not necessary to resolve the problem then before the court. In relation to the decision in *Elliott*, it is also pertinent to observe that it flowed from an arbitration in which the power of the arbiter was to award interest if the claimant was entitled to it. That is, as the court observed, if the claimant had a right to it by the application to the circumstances of the relevant law. In *Elliott*, the arbiter did not have a general discretion to award interest from such date and at such rates as he saw fit. Such a general discretion in relation to interest is found in clause 41B of JCT 98, as amended by the Scottish Building Contract. The introduction of such a clause is necessary as, at common law, an arbiter in Scotland does not have power to award interest from a date prior to that of his award. As arbitration remains a popular forum for the resolution of construction disputes, the existence of provisions such as clause 41B.6 will, in fact, go a long way to removing the uncertainty that remains after the decision in *Elliott*.

### 9.6.2 The Late Payment of Commercial Debts (Interest) Act 1998

It is interesting to note that the Court of Session in *Elliott* stated that it was a matter of concern that in modern commercial contexts the law did not, in general, allow for interest to run on debts from a date earlier than judicial demand (i.e. the date of service of a writ) and that reform of the law on interest on debts was a matter for government. At or about the time of the decision in *Elliott*, this was a subject upon which the government had been consulting and that process resulted in the enactment of the Late Payment of Commercial Debts (Interest) Act 1998 which came into force on 1 November 1998.

The Late Payment of Commercial Debts (Interest) Act 1998 ('the 1998 Act') was introduced with a view to encouraging purchasers to pay on time and to compensate suppliers where late payment persisted. The right to claim interest is to compensate suppliers for not being able to make use of the money owed to them and to cover the cost of increased borrowing resulting from late payment. The 1998 Act provides suppliers with a statutory right to interest on late payments. The rate of interest has been prescribed at 8% over the base rate of the Bank of England. The 1998 Act operates by implying a term into contracts to which it applies to the extent that any qualifying debt carries interest at the prescribed rate. A 'qualifying debt' is simply one where an obligation to make payment of the contract price arises under a contract to which the 1998 Act applies i.e. a contract for the supply of goods and/or services where both the purchaser and supplier are acting in the course of a business. The interest to which the supplier is entitled is simple interest.

The intention is that the Act will be implemented in three stages, the first of which is already in place.

At present, what are termed as 'small business' suppliers have the right to claim interest at the statutory rate against large businesses and United Kingdom Public Authorities. For the purposes of the 1998 Act a 'small business' is defined as one which has 50 or fewer full-time employees or part-time equivalents. A 'large business' is one with more than 50 full-time employees or part-time equivalents. A 'United Kingdom Public Authority' is defined at length in the commencement order for the 1998 Act but, in general, it means any emanation of the State.

The Government has indicated that after two years of operating the first phase, the right to claim interest will be extended to entitle all small business suppliers to claim interest at the statutory rate against all other businesses (i.e. including small businesses) and United Kingdom Public Authorities.

After a further two years, the intention is that the statutory right to claim interest will be further, and fully, extended. At that juncture, all businesses and United Kingdom Public Authorities will have the right to claim interest against all other businesses and United Kingdom Public Authorities.

A supplier is free to decide whether or not to claim interest. The statutory right is not compulsory. The right to claim interest arises when a payment is late. A payment is late when it is not made by the 'relevant day'. The relevant day is the date agreed for payment or, in the event that no such date has been agreed, the last day of the period of 30 days beginning with the later of

(a) the day of the supply/performance; or
(b) the date of notice to the purchaser of the amount of the debt - in practice, the invoice date.

Different rules exist where the contract requires advance payment. These are dealt with in s.11 of the 1998 Act. The principle is that the 1998 Act does not give a right to claim interest unless and until at least some of the goods have been delivered or part of the service performed. In essence, the s. 11 provisions allow for the right to claim interest 30 days after delivery/ performance.

Once the payment is late interest runs at the prescribed rate from the day after the relevant day until the principal sum is extinguished by payment. Unless the supplier accepts a payment on other terms, any payment received goes first to extinguish or reduce the accrued interest. A claim for interest is made by the supplier informing the purchaser, once the payment is late, that they are claiming interest. Notification can be in any fashion but it would appear prudent to make such a claim in writing. A claim for interest need not be made immediately. The ordinary rules of prescription will apply. In this regard, see Section 8.9.

It is, of course, common to find standard terms and conditions providing for interest to run on late payment. In recognition of that, the 1998 Act provides that, where arrangements have already been made, the statutory right to interest will not apply. To prevent purchasers abusing their right to agree arrangements with a supplier, any contractual remedy must be what is termed a 'substantial remedy'. This term is defined by s.9 of the 1998 Act. A remedy for late payment is 'substantial' if it is

(a) sufficient to compensate the supplier for the cost of late payment or to deter late payment; and
(b) it is fair and reasonable to allow the remedy to oust or vary the statutory interest that would otherwise apply.

In determining whether or not a remedy satisfies the fair and reasonable test, regard is to be had to

- the benefits of commercial certainty;
- the relative strength of bargaining power between the parties;
- whether the term was imposed by one party to the detriment of the other; and
- whether the supplier received an inducement for agreeing to the term.

If the contractual remedy is not a substantial remedy it is void. It will be interesting to see if any amendment is made to JCT 98 to increase the rate of interest payable under clause 30. This is considered further below.

### 9.6.3 Interest under JCT 98

There were also developments in relation to the entitlement to interest under JCT 80 contracts shortly after the decision of the Inner House in *Elliott*. Amendment 18 to JCT 98, issued in April 1998, introduced amendments to clause 30 giving a contractor, for the first time, a right to receive simple interest from the employer in certain defined circumstances.

The entitlement to interest in respect of interim certificates is dealt with by clause 30.1.1. If the employer fails properly to pay the amount, or any part thereof, due to the contractor under the conditions by the final date for its payment the employer is obliged to pay to the contractor, in addition to the amount not properly paid, simple interest thereon for the period until such payment is made. Payment of such simple interest is treated as a debt due to the contractor by the employer. The rate of interest payable is 5% over the base rate of the Bank of England current at the date the payment by the employer became overdue. This rate should be contrasted with the statutory rate provided for by the 1998 Act. It will be interesting to see whether issue is taken with the validity of this provision. It is conceivable that it could be argued that it is not a substantial remedy in the context of s.9 of the 1998 Act.

Clause 30.8.5 makes similar provisions in respect of sums due under a final certificate, whether those sums are due to the employer or the contractor. In each case, any payment of simple interest under the clause in question shall not, in any circumstances, be construed as a waiver by either party of their right to proper payment of the principal amount due.

### 9.6.4 Interest on damages

In terms of the Interest on Damages (Scotland) 1958, as amended by the Interest on Damages (Scotland) Act 1971, where a court grants decree for payment by any party of a sum of money as damages, the court's order may include provision for payment by that party of interest on the whole or any part of the amount of damages for the whole, or any part, of any period between the date when the right of action arose and the date of the court's order. The court also has a discretion as to the rate or rates at which such interest is to be paid. The mere fact that a right of action arose on a particular date prior to decree does not, of itself, justify an award of interest from that date, see *James Buchanan & Co Ltd* v. *Stewart Cameron (Drymen) Limited* (1973).

In *MacRae* v. *Reed and Malik Ltd* (1961) the Inner House of the Court of Session stated that the discretion conferred upon the court by the 1958 Act must be exercised on a selective and discriminating basis, and that the exercise of that discretion was open to review on the question as to whether the circumstances of the case warranted the course taken. They also held that interest from a date earlier than the date of decree could be allowed only on damages awarded for loss suffered before the date of decree and where such loss could be definitely ascertained.

## 9.7 Interdict

Where it can be demonstrated by a party that a legal wrong is continuing or that they are reasonably apprehensive that such a wrong will be committed, they are entitled to seek interdict against the wrong, see *Hay's Trustees* v. *Young* (1877). If the wrong has been completed, and it cannot be contended that there is a likelihood of it recurring, interdict will not be granted, see *Earl of Crawford* v. *Paton* (1911).

Both permanent and interim interdict can be granted in either the Court of Session or the Sheriff Court. In practice, few actions in which interdict is sought proceed beyond the interim interdict stage. The grant or refusal of interim interdict is often determinative of the issue between the parties.

Assuming the pursuer can satisfy the court that they have title and interest to bring the action and that they are confronted by, or threatened with, a wrong on the part of the defender, interim interdict will still only be granted if the balance of convenience favours the pursuer. To meet this test, the pursuer must demonstrate a cogent need for interim interdict, see *Deane* v. *Lothian Regional Council* (1986).

A detailed examination of the law of interdict is beyond the scope of this book, however, in the context of building contracts its availability as a remedy should not be overlooked. The remedy is available should there be a continuing, or reasonably anticipated, breach of contract.

## 9.8 *Withholding payment*

### 9.8.1 General

Having examined the issue of payment in Chapter 7, it is appropriate to consider the remedies that are open to a party under a building contract who is, on the face of it, obliged to make payment, but has reasons for not doing so. In Scotland, there exist two distinct and separate remedies, namely, retention and compensation. These are frequently confused. The term 'set-off' is often used in place of compensation. It is also to be found in the 1996 Act, see s.110(2)(b). In this section we will consider retention and compensation in the context of payment obligations. It should, however, be noted that retention applies not only to obligations to pay but also to all other obligations incumbent upon a party under a contract. The wider application of retention is considered below in section 9.9.1. Finally, we will consider the statutory right of withholding payment as contained within s.111 of the 1996 Act.

### 9.8.2 Retention

The principle of retention is, perhaps, best illustrated by the opinion of Lord Shand in *Macbride* v. *Hamilton* (1875) in which he stated:

> '[I]n cases of mutual contract a party in defence is entitled to plead and maintain claims in reduction or extinction of a sum due under his obligation where such claims arise from the failure of the pursuer to fulfil his part of the contract.'

For retention to operate, both claims must arise from the one contract. Retention, when considered in the context of withholding payment, should not be confused with compensation. Retention is, in effect, a form of security, whereas compensation extinguishes a debt, in whole or in part.

Retention has long since been a favoured remedy in building contract disputes, see for example *Johnston* v. *Robertson* (1861). In that case, the employers were entitled to plead in defence a claim for liquidated damages for non-completion against the contractors' claim for the balance of the contract price and payment for extra works.

In Scotland, there was authority which suggested that the general rule in respect of retention may not apply to the case of a building contract which contained provision for payment by instalments, it being doubted whether the employer had any right to withhold payment of an instalment by virtue of a claim against the contractor, see *Field & Allan* v. *Gordon* (1872). That position has, however, been accepted to be incorrect. Unless it is shown in clear and unequivocal words that the parties had agreed in the contract that the common law right of retention was to be excluded, that right would be available in respect of breaches of contract, see *Redpath Dorman Long Ltd* v. *Cummins Engine Co Ltd* (1981). Retention is considered further in section 9.9.1.

### 9.8.3 Compensation

The essence of compensation is that sums are due at the same time by parties to each other. Where each party owes the other a sum of money, compensation can operate to extinguish, or part extinguish, the debts. Certain prerequisites must be satisfied. Firstly, the debts must be due at the same time. A debt that is due at a future date cannot be set-off against one that is presently due, see *Paul & Thain* v. *Royal Bank* (1869). Secondly, each debt must be what is termed 'liquid'. A liquid debt is one that is for a readily ascertainable amount and is not disputed. A claim for damages is not a liquid debt, see *National Exchange Co* v. *Drew* (1855). There must also be what is termed *concursus debiti et crediti,* that is that each party must owe money and be owed money in the same capacity. An example of this is the case of *Stuart* v. *Stuart* (1869) in which it was held that the defender, as an individual, could not plead in compensation certain alleged counterclaims competent to him as his father's executor.

In Scotland, there has been legislation governing compensation for over four hundred years, see the Compensation Act 1592.

### 9.8.4 Withholding payment under the 1996 Act

In construction contracts, as defined by the 1996 Act (see Section 1.2.2) the circumstances in which payment of a sum due under the contract may be withheld are clearly defined. A party to a construction contract may not withhold payment after the final date for payment of a sum due under the contract unless they have given an effective notice of intention to withhold payment. To be effective, such a notice must specify the amount proposed to be withheld and the ground for withholding payment. If there is more than one ground, each ground and the amount to be withheld in relation to it must be specified. That notice must be given not later than the prescribed period before the final date for payment. Parties are free to agree what that prescribed period is to be, but if they do not the Scheme for Construction Contracts applies.

Paragraph 10 of Part II of the Schedule to the Scheme provides that any notice of intention to withhold payment shall be given not later than seven days before the final date for payment under the contract. The notice of intention to withhold payment can form part of the notice that is required under section 110 (2) of the 1996 Act, being the notice which specifies the amount of the payment made or proposed to be made, and the basis upon which that amount is calculated.

Clause 30.1.1.4 of JCT 98 contains provisions regarding the giving of a notice of withholding in relation to a sum which is due under an interim certificate. Not later than five days before the final date for payment of an amount due under an interim certificate, the employer may give a written notice to the contractor which specifies any amount proposed to be withheld and/or deducted from the due amount. The notice must also specify the ground or grounds for such withholding and/or deduction and the amount of withholding and/or deduction attributable to each ground. A similar provision exists in relation to final certificates under JCT 98, namely, clause 30.8.3.

## *9.9 Suspending performance*

### 9.9.1 General

The principle of retention, considered above in section 9.8.2 in the context of withholding payment, has a wider application. That wider application emanates from what is known as the mutuality principle. It is perhaps best shown in the opinion of Lord Benholme in the building contract case of *Johnston* v. *Robertson* (1861). Lord Benholme stated that:

> 'One party to a mutual contract, in which there are mutual stipulations, cannot insist on having his claim under the contract satisfied unless he is prepared to satisfy the corresponding and contemporaneous claim of the other party to the contract.'

Accordingly, where the common law right of retention is open to a party, such as a contractor, they are entitled to suspend performance when confronted by an employer who refuses to pay. It should be noted that it is possible to contract out of the common law right of retention, see *Redpath*

*Dorman Long Ltd* v. *Cummins Engine Co Ltd* (1981). That can only be achieved by the use of clear and unequivocal words in the parties' contract.

### 9.9.2 Suspension of performance under the 1996 Act

The 1996 Act provides a right to suspend performance for non-payment where a sum due under a construction contract is not paid in full by the final date for payment and no effective notice to withhold payment has been given. In those circumstances, the person to whom the sum is due is entitled to suspend performance of their obligations under the contract to the party by whom payment ought to have been made. This right is without prejudice to any other right or remedy open to the party entitled to payment. This would allow them to raise separate proceedings for payment, should they so wish. The right to suspend performance does not deprive the entitled party of any other rights competent to them. The right to suspend performance under the 1996 Act only arises in the event of non-payment.

The right may not be exercised without first giving to the party in default at least seven days' notice of intention to suspend performance. The notice must state the ground or grounds upon which it is intended to suspend performance. The contract can stipulate that a period of in excess of seven days notice of intention to suspend performance must be given. In practice, employers under a main contract (and main contractors in a sub-contract) will insist upon a greater period of notice. The right to suspend performance ceases when the party in default makes payment in full of the amount due. Any period of suspension of performance is disregarded in computing the time taken by the party to complete the works. Not only does this apply to the party exercising the right to suspend, but also to any affected third party.

### 9.9.3 Suspension under JCT 98

The contractors' statutory right of suspension is provided for by clause 30.1.4 of JCT 98. A written notice of intention to suspend must be given to the employer, with a copy to the architect. If the failure to pay continues for seven days after that notice is given the contractor may suspend performance of their obligations under the contract to the employer until payment in full occurs. A suspension under clause 30.1.4 is not a default by the contractor under clause 27.2.1.1, nor is it a failure to proceed regularly and diligently with the works, another contractor default, under clause 27.2.1.2.

By virtue of clause 25.4.18 a delay arising from a suspension by the contractor of the performance of their obligations under the contract pursuant to clause 30.1.4 is a relevant event which may entitle the contractor to an extension of time, see Section 5.5.2 above. Further, such a suspension is also a matter materially affecting the regular progress of the works under clause 26.2.10 which may entitle the contractor to recover direct loss and/or expense, considered above in Section 7.3.

## 9.10 Delictual claims

### 9.10.1 General

One party can owe a duty to another in the absence of a contractual relationship. In the context of building contracts, for example, a sub-contractor owes certain duties to the employer. See *British Telecommuncations plc* v. *James Thomson & Sons (Engineers) Ltd* (1999).

Liability in delict in construction projects is most likely to arise under the law of negligence or the law of nuisance, although claims may also arise in relation to breach of statutory duty.

### 9.10.2 Losses recoverable

Broadly speaking, in order to establish a claim in negligence the pursuer must show

- that the defender owed the pursuer a duty of care in respect of the type of loss in question;
- that this duty was breached;
- that the breach of duty caused the pursuer's loss; and
- that the loss is not too remote.

It should be noted that the precise rules on remoteness of damage differ between claims based on breach of contract and those based on delict. See *Koufos* v. *C Czarnikow Ltd* (1967). In delict, the losses recoverable are those reasonably foreseeable to the defender at the time of the negligent act. See *Allen* v. *Barclay* (1864). In breach of contract cases, on the other hand, the losses recoverable are those foreseeable at the time the contract is entered into, see section 9.4.2. The reasoning behind this distinction is that in a contract there is the opportunity for one party to obtain protection against a particular type of potential loss by directing the other party's attention to it before the contract is made. In cases arising out of delict there is no such opportunity.

### 9.10.3 Economic loss

As a general rule, the losses claimed in delict must not be too remote. This means that damages for personal injury, death and loss of or physical damage to property (and economic loss flowing from such loss of or physical damage to property) arising from a breach of duty would normally be recoverable. However, the right to recover economic loss in the absence of physical damage is a particularly problematic area. A detailed examination of the issue is beyond the scope of this book, but the following is a brief overview.

A convenient starting point, which illustrates the type of situation in which matters of this nature arise, is the decision of the House of Lords in the Scottish case of *Junior Books Ltd* v. *The Veitchi Co Ltd* (1982). In this case Junior Books owned a factory. They entered into a contract with builders for, amongst other things, the laying of flooring in the factory's production area. The builders sub-contracted this work to Veitchi, who were specialist flooring contractors. Junior Books subsequently raised an action against Veitchi, seeking damages for loss allegedly sustained as a result of their negligent workmanship. This loss included the cost of

(a) replacing the floor surface, allegedly defectively laid;
(b) storing goods and moving machinery during the period of replacement;
(c) paying wages to employees unable to work during this period; and
(d) fixed overheads which would produce no return during this time.

Junior Books also claimed for loss of profit sustained by the temporary closure of the business. They argued that Veitchi, as specialists, knew what products were required; were alone responsible for the composition and construction of the flooring; must have known that Junior Books had relied upon their skill and experience; and must be taken to have known that if they did the work negligently, Junior Books would suffer economic loss in requiring to expend money to remedy the resulting defects. Junior Books did not argue that actual or prospective danger to persons or property arose from the state of the flooring. If they had done so there would have been a duty of care under the principles laid down by *Donoghue* v. *Stevenson* (1932).

Veitchi argued that the case was irrelevant in law. They contended that the law did not make them liable in delict for the cost of replacing the floor or for economic or financial loss consequent upon that replacement. They argued that while they were under a duty of care to prevent harm being done to property or persons by their faulty work (in accordance with *Donoghue*), they had no duty of care to avoid such faults being present in the work itself. They argued that for the court to hold otherwise would extend the duty of care owed by manufacturers and others far beyond the limits to which the courts had previously extended them; and that a manufacturer's duty not to make a defective product set a standard of care which was much less easily ascertained than that for a duty not to make a dangerous product.

The Inner House of the Court of Session and the House of Lords rejected that argument. They held that there was the enough proximity between Junior Books and Veitchi so as to give rise to the relevant duty of care relied on by Junior Books. Further, they held that there were no considerations in this particular case to negative, restrict or limit that duty of care. Pure economic loss was the sort of loss which Veitchi, standing in the relationship to Junior Books which they did, ought reasonably to have anticipated as likely to occur if their workmanship was faulty.

The law in England in relation to economic loss now rests with the decision of the House of Lords in *Murphy* v. *Brentwood DC* (1991). In *Murphy* it was held that the defendants, who had negligently approved plans that contained erroneous calculations submitted by the builders constructing the plaintiff's house, owed no duty of care to the plaintiffs. The consequence of the plans being incorrect was that the plaintiff, upon selling the house, was unable to obtain the full market value. In this case it was clear, according to the court, that there was no proximity between the plaintiff and the defendants. In essence it could not be said that the defendants had assumed any responsibility to the plaintiff in respect of the plans which they approved.

Further policy considerations are evident in their lordships' decision, namely the 'floodgates' argument and the fear that had a duty been imposed the court would have introduced a transmissible warranty of quality into property transactions which was a legislative matter for Parliament. A further consequence of the decision in *Murphy* was the restriction of *Junior Books Ltd* to its own special facts, mirroring the court's reluctance in previous cases, such as *D & F Estates Ltd* v. *Church Commissioners for England* (1989), to apply the decision in *Junior Books Ltd.*

As we have already seen in *Murphy,* which was also followed in the subsequent case of *Department of the Environment* v. *Thomas Bates & Son* (1990), foreseeability of harm based on *Donoghue* principles will not of itself generally be sufficient to impose a duty of care for economic loss. Rather, the courts now require to determine whether there was sufficient proximity between the parties (see *Murphy*) to justify the imposition of a duty of care and also whether it is fair, just and reasonable to impose a duty, see *Caparo Industries plc* v. *Dickman* (1990).

The law in England in relation to economic loss was reviewed by the House of Lords in *White and Another* v. *Jones and Others* (1995). Whilst the case addresses the issue of whether a solicitor, who negligently drew up a will, owed a duty of care to disappointed prospective beneficiaries, it is submitted that the principles enunciated by the court are of general application. The court expounded the view that proximity was to be assessed by considering whether it could be said that the party causing the loss assumed responsibility in whatever form to the party suffering that loss. The court also reaffirmed their commitment to allowing recovery by analogous extension only. That is, the court will look to determine whether recovery has been permitted in similar situations before allowing recovery in the case under consideration. The law lords in *White and Another* reiterated their opposition to any *carte blanche* extension of the law in relation to economic loss. In essence the law will only be allowed to develop by increment rather than by quantum leap.

Notwithstanding the above, it must be borne in mind that *Junior Books,* being a House of Lords decision in a Scottish case, is still binding upon Scottish courts. It has not been overruled by any of the subsequent House of Lords decisions in English cases. Instead, its applicability has been stated (both by the courts and by legal commentators) to be confined only to the very limited circumstances which pertained to the facts of that case. However there have been several decisions of the Scottish courts which have at least indicated that, while the House of Lords decisions in *Murphy* and *D & F Estates* will be of very high persuasive authority to a Scottish court, it is perhaps rather premature to assume that such cases will be followed unquestioningly by Scottish courts or that, as some commentators would suggest, *Junior Books* is 'dead and buried', see *Parkhead Housing Association Ltd* v. *Phoenix Preservation Ltd* (1990) and *Scott Lithgow Ltd* v. *GEC Electrical Projects Ltd* (1992).

To conclude this brief overview of economic loss it is perhaps worth considering one of the major policy restrictions often cited as the principle reason for imposing restrictions on the recovery of economic loss, namely the 'floodgates' argument. The floodgates argument should not be misunderstood as being a reflection of the courts' unwillingness to countenance a multitude of claims against one party. Rather the floodgates argument is the courts' unwillingness to allow liability in an indeterminate amount for an indeterminate time to an indeterminate class. The courts' formulation of the test of proximity will, inevitably, filter out those claims where liability is indeterminate.

# Chapter 10
# Sub-contractors and Suppliers

## 10.1 *Introduction*

On any large construction project it is not uncommon for the majority of the works to be performed by sub-contractors. Indeed it is not unheard of for all the works to be sub-contracted by the main contractor. In turn many sub-contractors will themselves engage sub-sub-contractors. The standard forms of sub-contract produced by the SBCC for use in Scotland are listed in Sections 1.5.2 and 1.5.3. The principles regarding the formation and interpretation of a building contract, considered in Chapter 2, apply equally to sub-contracts. This chapter will deal with the types of sub-contractor and the relationship between employer, main contractor and sub-contractor, and finally will outline some of the issues which frequently arise in practice.

## 10.2 *Nominated and domestic sub-contractors*

Under the umbrella of the JCT forms of main contract, sub-contractors are typically either domestic or nominated. A nominated sub-contractor will submit his quotation to the architect or quantity surveyor and in turn be nominated on behalf of the employer if successful. A domestic sub-contractor is usually invited by the main contractor to tender competitively. Historically main contractors have always exhibited a preference for domestic sub-contracts. The SBCC has published the Sub-Contract DOM/A/Scot (September 1997 Edition - August 1998 Revision) as the standard form for domestic sub-contracts in Scotland where the main contract is governed by JCT 80. It is anticipated that this will be updated to tie in with JCT 98 and that a separate form will be published in respect of domestic sub-contractors who assume a design responsibility.

What are the important differences between nominated and domestic sub-contracts? An important difference exists between the respective payment provisions. Under clause 35.13.1.1 of JCT 98 the architect is obliged to identify that portion of any interim valuation which relates to work carried out by nominated sub-contractors. Thereafter, the architect must inform each nominated sub-contractor of the amount of any interim or final payment allocated to their work. Prior to any further certification in favour of the main contractor, the main contractor is obliged to provide the architect with reasonable proof of payment to the sub-contractor. No such provisions exist under JCT 98 in respect of payments to domestic sub-contractors.

In certain circumstances, it has been held possible for an aggrieved sub-contractor to recover damages from the employer in respect of a failure to observe this provision which resulted in loss to the sub-contractor, see *Pointer Ltd* v. *Inverclyde DC* (1990). Prior to the implementation of s. 113 of the 1996 Act, which outlaws certain forms of 'pay when paid' clauses, domestic sub-contractors had little scope for identifying, with any degree of precision, what sums, if any, had been paid by an employer to a main contractor in respect of the sub-contract works. Even with the assistance of the recovery of documents through the courts, the position could often remain blurred, particularly when a large proportion of the main contract works had been carried out by the main contractor rather than by sub-contractors.

The second important difference between nominated and domestic sub-contracts is the ability of the nominated sub-contractor under JCT 98 to secure payment direct from the employer. In the absence of an express term in the main contract enabling an employer to pay a sub-contractor direct the employer should not do so as his obligation to pay the main contractor for the sub-contract works is not satisfied by making payment to the sub-contractor. Any direct payment provision should not only confer a right on the employer to make a direct payment to a sub-contractor in certain defined circumstances, but must also allow the employer to deduct an equivalent amount from sums otherwise due to the main contractor. Clause 35.13.5 of JCT 98 provides that where the contractor has failed to provide reasonable proof of payment to a nominated sub-contractor of the amount included in an interim certificate, the architect shall issue a certificate to that effect; the amount of any future payment otherwise due to the contractor shall be reduced by the amount due to nominated sub-contractors which the contractor has failed to pay, and the employer shall himself pay the same to the nominated sub-contractors concerned. The obligation on the part of the employer to make direct payment will, however, only arise if the employer has entered into the direct agreement, NSC/W/Scot, with the nominated sub-contractor.

No direct payment can be made if, at the date when the deduction and payment to the nominated sub-contractor would otherwise be made, the main contractor has become bankrupt, has become 'apparently insolvent' or has had a winding up order made. This leaves open the question whether the direct payment provisions can be operated upon the appointment of an administrator or receiver to the main contractor.

JCT 98 does allow direct payment to domestic sub-contractors in certain circumstances. Clause 27.5.2.2 (as amended by the Scottish Building Contract) provides that in the event of determination of the main contractor's employment, and except where the determination occurs due to an insolvency event of the type specified in clause 27.3.2 (bankruptcy, winding up, receivership etc. of the contractor), the employer may make direct payments to sub-contractors and suppliers (whether nominated or not). It should be noted that, in contrast to the mechanism contained in clause 35.13, the employer has a discretion, not an obligation, under clause 27.5.2.2 to make a direct payment.

Nomination has the advantage to employers of allowing them to specify who actually performs the relevant sub-contract works. This may be important in the case of specialist sub-contract works.

Given that much, if not all, of the work performed by nominated sub-contractors will be of a specialist nature, what is the liability of the main contractor to the employer in the event of default by the nominated sub-contractor? Where there has been delay on the part of the nominated sub-contractor (or nominated supplier), which the main contractor has taken reasonable steps to reduce, clause 25.4.7 of JCT 98 provides that this is a relevant event for the purposes of the main contractor securing an extension of time.

Clause 35.21 provides that the main contractor has no liability to the employer in four specified situations.

- The main contractor does not have a responsibility for the design of any nominated sub-contract works insofar as they were designed by the nominated sub-contractor.
- The main contractor does not have responsibility for the selection of the kinds of materials and goods which have been selected by the nominated sub-contractor.
- The main contractor has no responsibility for the satisfaction of any performance specification or requirement insofar as that is included in the nominated sub-contract works.
- The main contractor has no responsibility for the provision of information by the sub-contractor in reasonable time in order that the architect can comply with the relevant provisions of the main contract.

Sir Michael Latham, in his 1994 report *Constructing the Team*, recorded that the Chartered Institute of Purchasing and Supply had described nomination as a 'contradiction in terms' and had recommended its abolition. He also noted that the New Engineering Contract made no provision for nomination. Nonetheless, the nomination of sub-contractors remains an integral part of the JCT 98 contracts.

Mention should also be made of 'named' sub-contractors. Clause 19.3 of JCT 98 provides that, in certain circumstances, work must be carried out by one of a number of persons named in a list which is either in or annexed to the contract bills. The work in question will have been priced by the main contractor, and the selection of the person to carry out the work is at the sole discretion of the main contractor. This procedure has the benefit for the employer that certain specialist work will be carried out by suitably experienced sub-contractors, whilst not involving the employer with the complications that can be associated with nomination.

## 10.3 *Privity of contract*

### 10.3.1 General

Ordinarily there is no direct contractual relationship between the employer and the sub-contractor, and the individual contracts which make up the contractual chain between sub-contractor and employer are (subject to collateral warranties) enforceable only by the parties to such contracts. This principle is known as privity of contract.

The law would, however, in certain circumstances, permit an employer to sue a supplier direct, should the supplier have given certain assurances or warranties as to, for example, the fitness for purpose of the supplier's product, notwithstanding the fact that the contract for the sale of the product was with the contractor appointed by the employer, see *Shanklin Pier Ltd* v. *Detel Products Ltd* (1951) and *British Workman's and General Assurance Co* v. *Wilkinson* (1900).

Another exception is where appropriate rights are assigned by a main contractor to a sub-contractor, see *Constant* v. *Kincaid & Co* (1902). Assignation is considered below in Chapter 11. In the absence of a direct contractual relationship, or an assignation of rights, neither employer nor sub-contractor can sue the other under contract. A practical example of this is that, in the absence of the direct agreement, NSC/W/Scot, an employer and nominated sub-contractor would not be able to enforce any rights against the other, for example direct payments (see above).

### 10.3.2 *Jus quaesitum tertio*

The above statement on privity of contract is, however, qualified where a *jus quaesitum tertio* has been created by the contract. The creation of such a right will give a third party (the 'tertius') a right to sue under the contract, notwithstanding that he is not a party to it. In order to create the right, the contract must expressly, or by implication, confer a benefit on the tertius or a class of persons of which the tertius is a member. Unless an intention on the part of the contracting parties to create a *jus quaesitum tertio* in favour of a third party is expressed or can be inferred from the terms of the contract, no such right will be created. See *Scott Lithgow Ltd* v. *GEC Electrical Projects Ltd* (1992) and *Strathford East Kilbride Ltd* v. *HLM Design Ltd* (1997).

### 10.3.3 Delict

Prior to certain case law in the late 1980s, and in particular, *D & F Estates Ltd* v. *Church Commissioners for England* (1989) and *Murphy* v. *Brentwood DC* (1991) it had been understood that an employer could sue a sub-contractor direct under delict and recover economic loss, see *Junior Books Ltd* v. *The Veitchi Co Ltd* (1982). Collateral warranties emerged as a result of *D & F*

*Estates Ltd* and *Murphy*. Market forces dictated that any perceived vacuum in the law of negligence be filled by the law of contract. Developers, owners and funders of large commercial developments need the ability to sue professional team members and/or specialist sub-contractors.

This branch of the law of negligence has nevertheless continued to develop in both Scotland and England. In *White and Another* v. *Jones and Others* (1995) the House of Lords in an English appeal decided by a majority that a solicitor owed a duty of care to beneficiaries under a will which had been negligently drawn up. In the Scottish case of *Scott Lithgow Ltd* Lord Clyde allowed to proceed to proof a case in which an employer sued a domestic sub-contractor for recovery of economic loss stemming from allegedly defective wiring. He held that nomination was not a necessary factor before a duty of care could arise but it was an important element where it did exist. Where it does exist it obviously serves to point towards the degree of proximity which is required for the employer to succeed.

It must be fair, just and reasonable for a duty of care to exist. This issue was recently addressed by the House of Lords in the Scottish case of *British Telecommunications plc* v. *James Thomson & Sons (Engineers) Ltd* (1999). In that case the employer sued sub-contractors in delict in respect of losses sustained as a consequence of a fire breaking out in the employer's premises for which the employer held the sub-contractors responsible. The sub-contractors had been engaged by the main contractor on the same terms and conditions of contract as those ruling between the employer and the main contractor. The insurance provisions in the main contract made it clear that damage caused in the way suggested by the employer was to be covered by an insurance policy which the employer was bound to take out. In short, the damage in question was one of the specified perils under the main contract. As such it was contended by the sub-contractors that it would not be fair, just or reasonable to impose a duty on them to avoid such damage. Their Lordships however attached significant weight to the fact that the insurance arrangements in the main contract afforded any nominated sub-contractor the benefit of a waiver by the relevant insurers of any right of subrogation which they may have against the nominated sub-contractor but no such provision existed for the benefit of domestic sub-contractors. As such the unanimous decision of the court was that it would be fair, just and reasonable to impose a duty of care on the domestic sub-contractors to the employer.

## 10.4 Relationship between main and sub-contracts

It is common for main contractors to attempt to incorporate by reference the terms of the main contract into the sub-contract. This practice of wholesale incorporation is not to be encouraged and frequently leads to disputes between the parties. The degree of incorporation can vary. Many main contractors attempt to incorporate their own programme into the sub-contract, see *Scottish Power plc* v. *Kvaerner Construction (Regions) Ltd* (1998).

The effect of incorporation of main contract terms was considered in

*Babcock Rosyth Defence Ltd* v. *Grootcon (UK) Ltd* (1998). In this case the sub-contractor raised an action against the main contractor. The main contract incorporated a modified form of the ICE Conditions of Contract (5th edition). The issue for the court was whether or not clause 66, the arbitration clause, formed part of the sub-contract. The main contractor maintained that the ICE 5th edition was incorporated into the sub-contract, subject to the express qualifications made and to its adaptation for practical effectiveness in the sub-contractual relationship. To that extent the main contractor's submissions did not go as far as those made in *Parklea Ltd* v. *W & J R Watson Ltd* (1988). In the latter case the court rejected the contention that the whole provisions of the main contract were to be incorporated *mutatis mutandis* into the sub-contract. In *Babcock Rosyth Defence Ltd* the defenders acknowledged that certain of the main contract provisions would have no place in the sub-contract. The judge, Lord Hamilton stated:

> 'When parties make reference to a set of conditions designed primarily for use in another contract but do not expressly adapt those conditions to meet the circumstances of their own relationship, it is often difficult to determine with confidence the contractual effect. Where, on the one hand, the circumstances demonstrate a plain common intention to incorporate terms, albeit expressed in language designed primarily for another purpose, the court will, where it is possible to do so without substantially rewriting the parties' bargain, give effect to the parties' plain common intention by incorporating terms subject to appropriate linguistic adaptation ... Where, on the other hand, the common intention is not plain or there are major difficulties about linguistic adaptation, the result will be otherwise ... Even in cases where incorporation subject to linguistic adaptation is possible and appropriate, there may yet remain a question as to the extent to which conditions are so incorporated.'

Lord Hamilton held that the parties had plainly intended that the ICE 5th Edition should apply to some extent, albeit with appropriate linguistic adaptation. He was not satisfied, however, that it was sufficiently clear that the parties intended to incorporate the arbitration clause into the sub-contract. To avoid ambiguity, therefore, parties should make it clear which particular main contract provisions are to be incorporated into the sub-contract and to what extent.

## 10.5 Main contractor's discount

It is common for sub-contracts to allow the main contractor a discount on the price of the sub-contract works. Unless the parties to a sub-contract make an express provision which connects the main contractor's ability to deduct discount with prompt payment, the courts are likely to view the discount as being no more than a reduction in the sub-contract price, see *Team Services plc* v. *Kier Management and Design Ltd* (1993).

## 10.6 Suppliers

Contracts of supply or sale are regulated by the Sale of Goods Act 1979, as amended by the Sale and Supply of Goods Act and Services 1994. In terms of s.14 of the 1979 Act as amended by s.1(1) of the 1994 Act, there is an implied term that the goods supplied under a contract of supply or sale are of 'satisfactory quality'. The quality of goods is deemed to include their state and condition and, in appropriate cases, their fitness for all the purposes for which goods of the kind in question are commonly supplied, their appearance and finish, their freedom from minor defects, their safety and their durability.

Clause 36 of JCT 98 deals with nominated suppliers. The architect is required to issue instructions for the purpose of nominating a supplier for any materials or goods in respect of which a prime cost sum is included in the contract bills or arises under clause 36.1. Main contractors who are about to enter into a contract with a nominated supplier should take particular care when examining any terms and conditions of business which the supplier may be endeavouring to introduce into the contract. The architect has the power to approve in writing any restriction, limitation or exclusions imposed by the supplier. Should he do so, the liability of the main contractor to the employer will be restricted, limited or excluded to the same extent. The main contractor has a right to refuse the nomination if the architect does not exercise this power of approval.

## 10.7 Retention of title clauses

Suppliers' terms and conditions commonly include a retention of title clause. The object of such a clause is to protect the supplier against the insolvency of its customer by delaying the passing of ownership of the goods in question to the customer until payment has been made. Otherwise, the ownership of the goods will normally transfer upon delivery (Sale of Goods Act 1979, s.17).

Section 19 of the Sale of Goods Act 1979 permits a seller to retain ownership of goods, notwithstanding delivery to the purchaser, in the event that the parties to the contract expressly provide that change of ownership is to be conditional. The most obvious condition will, of course, be as to payment of the price. Section 19 is a restatement of the position at common law and under the Sale of Goods Act 1893. The period between the mid 1970s and 1990 saw considerable litigation on the subject of retention of title clauses starting with *Aluminium Industrie Vaassen BV* v. *Romalpa Aluminium Ltd* (1976), and ending with *Armour* v. *Thyssen Edelstahlwerke AG* (1990). In the latter case the House of Lords held that 'all sums' retention of title clauses were effective in Scotland, as had been the case in England for some time. Prior to the decision by the House of Lords, the Scottish courts had restricted the applicability of retention of title clauses to the extent that they reserved title to the seller of goods in the event that the purchase price for *those* goods had not been paid. The courts had refused to give effect to retention of title clauses which purported to reserve title to the seller until all sums due by the purchaser to the seller, including sums due in respect of other goods, had been paid.

A retention of title clause will not protect the unpaid supplier in the event of the contract of sale being governed by s.25(1) of the Sale of Goods Act, i.e. where a third party has purchased the relevant goods in good faith and without notice of the retention of title clause. Section 25(1) provides that:

> 'Where a person having bought or agreed to buy goods obtains, with the consent of the seller, possession of the goods or the documents of title to the goods, the delivery or transfer by that person, or by a mercantile agent acting for him, of the goods or documents of title, under any sale, pledge, or other disposition thereof, to any person receiving the same in good faith and without notice of any lien or other right of the original seller in respect of the goods, has the same effect as if the person making the delivery or transfer were a mercantile agent in possession of the goods or documents of title with the consent of the owner.'

The application of s. 25 is illustrated by the case of *Archivent Sales and Developments Ltd* v *Strathclyde Regional Council* (1985). In that case the supplier delivered goods to site and payment was made by the employer to the main contractor. The main contractor failed, in turn, to make payment to the supplier, who sought to recover the goods from the employer on the basis of the retention of title clause in its contract with the contractor. The supplier's claim to ownership failed because the existence of the supplier's retention of title clause in the contract of sale to the main contractor had not been brought to the employer's attention.

There are a number of other practical difficulties in enforcing a retention of title clause, not least of which is that the clause cannot be founded upon in a question with the building owner where the materials have been incorporated into the structure, provided that there is no element of bad faith on the part of the building owner, see *Archivent Sales and Developments Ltd* v. *Strathclyde Regional Council* (1985). This arises from the principle of Scots law that all buildings and fixtures pass into the ownership of the party who has title to the ground upon which they are erected, see *Brand's Trustees* v. *Brand's Trustees* (1876). Whether an article attached to a structure (as distinct from remaining moveable property) becomes, by virtue of the principle of accession, a fixture and thus part of the structure is a matter of fact to be determined by the circumstances of the case, see *Scottish Discount Co. Ltd* v. *Blin* (1986).

On a further practical level, the sellers will need to identify their goods and, if necessary, distinguish them from other similar goods. A seller of identifiable items bearing serial or batch numbers is likely to enjoy greater success in enforcing a retention of title clause than a supplier of sand or bricks. In the latter situation (in the absence of an 'all sums' retention of title clause) even if the sellers can identify the bricks they supplied, that will not be enough unless they can connect particular quantities of unfixed brick with particular unpaid invoices.

## 10.8 Supply of goods by sub-contractors

A distinction falls to be drawn between materials supplied by a supplier and those supplied under a sub-contract, see *Thos Graham & Sons* v. *Glenrothes Development Corporation* (1986). As seen above title to goods or materials supplied by a supplier will normally pass to the contractor upon delivery to site, unless title has been retained by the supplier under a retention of title clause. In contrast, in the case of a sub-contract, i.e. a contract for the supply of goods and services, the common law applies and ownership passes when the goods or materials are fixed to the structure. See *Stirling County Council* v. *Official Liquidator of John Frame Ltd* (1951).

The conceptual difference between goods or materials supplied under a contract of sale and those supplied under a contract for the supply of goods and services is further illustrated by the way in which the Scottish Building Contract deals with the purchase of off-site materials. Clause 30.3 of JCT 98 provides that the architect in his discretion may include in the amount stated as due in an interim certificate the value of any materials or goods before delivery to the site. Such a provision is particularly useful in the case of major equipment which is being manufactured by a specialist supplier in his premises and which needs to be paid for prior to delivery to site. Clause 16.2 provides that where payment has been made of the value of the off-site materials as stated in the interim certificate, the materials shall become the property of the employer. However, this provision is not effective in transferring property in the materials in question to the employer under Scots law, as in Scotland ownership of goods supplied under a contract for goods and services cannot pass until, at the earliest, delivery.

It is for this reason that the Scottish Building Contract deletes clauses 16.2 and 30.3 of JCT 98 and substitutes a new clause 30.3 which allows the employer to enter into a separate contract for the purchase from the contractor or any sub-contractor of any materials or goods prior to delivery to site. This has the result of characterising the transaction as a contract of sale, and thus subject to the Sale of Goods Act 1979, which allows ownership to pass prior to delivery.

In terms of clause 4.15.2.2 of the SBCC Domestic Sub-Contract Conditions, where the value of unfixed materials or goods which have been delivered to or placed on or adjacent to the works has been included in any interim certificate under clause 30.2 of the main contract, and the amount properly due by the employer to the main contractor has been discharged, then such materials or goods become the property of the employer, and the sub-contractor agrees that they cannot deny that such materials or goods are and have become the property of the employer. Should the main contractor pay the sub-contractor for any such materials or goods prior to the employer having first discharged his obligation to pay the main contractor for same, then clause 4.15.2.3 provides that ownership will vest in the main contractor.

# Chapter 11
# Assignation, Delegation and Novation

## 11.1 *Introduction*

This chapter considers three separate methods of transferring rights and obligations, namely, assignation, delegation and novation. Assignation is a method of transfer of moveable property. It is a common feature of modern building projects for the employer's heritable interest in the project to be transferred by sale or lease. To provide the purchaser or lessee with an interest in the building contract, it is necessary to assign that interest. In the standard forms of building and engineering contract, specific, detailed provisions are made in respect of assignation. The specific provisions of JCT 98 in this regard are considered below at section 11.5.

All contracts contain both rights and obligations. In England, it is trite law that the obligations under a contract, or the burden, cannot be assigned without the consent of the party entitled to enforce those obligations. See *Linden Gardens Trust Ltd* v. *Lenesta Sludge Disposals Ltd and Others* (1993). The position in Scotland is different to that in England. It is a matter of some controversy in Scotland as to whether the contract as a whole is capable of being assigned. In Scotland, it would appear to be the case that both rights and obligations under a contract are capable of being assigned, provided that they do not involve an element of *delectus personae* (a specific choice of individual or firm). *Delectus personae* is considered below in section 11.4. The distinction is an important one. Whilst certain rights and obligations may involve elements of *delectus personae*, and thus not be assignable without consent, certain others may not and so may be assigned. An express provision in a building contract that prohibits assignation, or only permits assignation with consent, will override the common law. In those circumstances, *delectus personae* is of no relevance.

## 11.2 *Common law*

In the absence of any express provision of the contract governing assignation, the common law will apply.

No particular wording is required to constitute an assignation under Scots law, see *Carter* v. *McIntosh* (1862). To be effective an assignation of a right must be intimated to the person against whom the right may be enforced. For example, where a contractor assigns his right to receive payment of retention monies, the assignee, that is the party in whose favour the assignation is granted, must intimate such assignation to the employer. In the event of competing assignations these rank in priority in accordance with the date of their intimation, not the date of execution of the assignation.

The assignee under a contract is entitled to no greater benefit than the assignor. This is often expressed by way of the Latin maxim *assignatus utitur jure auctoris,* which means that the assignee can never be in a better position than the assignor. Thus any defences available to the debtor in respect of the claim by the assignor will also be available against a claim by the assignee.

## 11.3 Effect of assignation upon claims

Questions often arise in the construction industry as to the effect of an assignation upon claims. This has been considered by the House of Lords in the Scottish case *of GUS Property Management Ltd* v. *Littlewoods Mail Order Stores Ltd* (1982) and in the English cases of *Linden Gardens Trust Ltd* v. *Lenesta Sludge Disposals Ltd* (1993), *St Martin's Property Corporation Ltd and Another* v. *Sir Robert McAlpine Ltd* (1993) and *Darlington BC* v. *Wiltshier Northern Ltd* (1995).

In *GUS Property Management Ltd* a building in Glasgow, owned by a company named Rest Property Co. Ltd, was damaged in the course of building operations being carried out on a neighbouring property. Rest was a wholly owned subsidiary of a company which subsequently adopted a policy under which various properties were to be transferred to a newly-created wholly-owned subsidiary company, namely GUS. Rest transferred the building to GUS for a figure representing its book value. Rest assigned to GUS all claims competent to them arising out of the building operations. GUS raised an action of damages against the neighbouring proprietors and those involved in the building operations. The defenders argued that because the property had been transferred for its full book value, the assignor had not sustained any loss at all, and thus the assignee could not recover damages for any such loss.

The House of Lords, over-ruling the Court of Session, held that because the price had been fixed for internal accounting purposes such a defence was not sustainable and refused to allow the claim to fall into some kind of legal 'black hole'. This followed the earlier Scottish decision in *Gordon* v. *Davidson* (1864). Whilst *GUS Property Management Ltd* is a case dealing with a delictual claim, it is submitted that the principles set out in it are equally applicable to claims under contract. However that decision was based on a specific set of circumstances and, where the transfer is at arm's length rather than an inter-group transaction, it is not entirely clear whether the court would reject a defence to a claim by an assignee on the grounds that no loss has been suffered because the assignor (i.e. the original developer) sold for full value. The English courts have resisted such 'no loss' arguments, see *Darlington BC.*

The appeals in *Linden Gardens Trust Ltd* and *St Martin's Property Corporation Ltd and Another* were heard together by the House of Lords. In each case the contract was subject to the JCT 63 conditions. Clause 17 of JCT 63 (as with JCT 98) prohibited assignation of the contract by the employer without the contractor's written consent. The House of Lords held that, on a true construction of the contracts, the wording of clause 17 prohibited assignation by the employer, without the contractor's consent, of the benefit of the contract and the assignation of any cause of action. However, it was also held that the original employer was entitled to recover from the contractor the loss sustained by the purchaser. As the contract was expressly not assignable without the contractor's consent, the House of Lords held that the employer and the contractor should be treated as having contracted on the basis that the employer would be entitled to enforce his contractual rights against the contractor for the benefit of third parties who would suffer from defective performance.

Whilst the decisions in *Linden Gardens Trust Ltd* and *St Martin's Property Corporation Ltd and Another* are of considerable interest, the extent to which they are, truly, applicable in Scotland remains to be seen. The reason for this is that the underlying premise in each case was that it was impossible to assign 'the contract' under English law. That may not be the position in Scotland. Nevertheless, the Court of Session has applied the decision of the House of Lords in *Linden Gardens Trust Ltd*, see *Steels Aviation Services* v. *H Allan & Son Ltd* (1996).

## 11.4 Delectus personae

A contract which involves *delectus personae* is one where a party to the contract has entered into it in reliance upon certain qualities possessed by the other. In such circumstances, the contract cannot be performed by a third party and, consequently, the obligation to perform cannot be assigned without consent, see *Anderson* v. *Hamilton & Co* (1875). Authorities in relation to delegation (see section 11.6) may assist in identifying whether or not a contract contains an element of *delectus personae*. If the contract does not place reliance upon a special skill of one of the parties, or is for the provision of an item of a certain standard specified in the contract, no element of *delectus personae* exists, see *Cole* v. *Handasyde & Co* (1909). Even if the contract involves *delectus personae,* and is thus not assignable, certain rights arising out of that contract may be assignable. For example, an accrued right to payment of a sum of money under the contract may be assigned, notwithstanding that the contract itself cannot, see *International Fibre Syndicate Ltd* v. *Dawson* (1901).

## 11.5 Assignation under JCT 98

The assignation provisions of JCT 98, as amended by the Scottish Building Contract, are to be found in clause 19. Neither the employer nor the contractor may assign the contract without the written consent of the other.

Clause 19.1.2, as amended by the Scottish Building Contract, also provides (if stated in the Abstract of Conditions to apply) for assignation by the employer of certain benefits after practical completion. Where the clause is stated to apply, in the event of the employer alienating by sale or lease, or otherwise disposing of his interest in the contract works, he may at any time after the issue of the certificate of practical completion assign to the party acquiring his interest in the works his right, title and interest to bring proceedings, in his name as employer, to enforce any of the rights of the employer arising under or by reason of breach of the contract. Such proceedings can be by way of arbitration or court proceedings.

This provision, which was introduced in 1987, recognises the practice that certain employers will transfer their interest in property once practical completion has been certified. Should it apply, however, the provisions of clause 19.1.2 are of limited use. It is only the employer's right, title and interest to bring proceedings which may be assigned, and there may be occasions where the employer wishes to assign prior to the practical completion. The wording is a little odd, given that an assignation of rights under the building contract would, *ipso facto*, confer the right to raise proceedings to enforce such rights, and the such proceedings would normally be brought in the name of the assignee. There appear to be no reported cases applying the provisions of clause 19.1.2.

## 11.6 Delegation

The delegation of building work, through the use of sub-contractors, is commonplace within the construction industry. Although delegation raises similar issues to assignation, the two concepts should not be confused. In the case of assignation, it is the assignee who is bound, as opposed to the original contracting party. With delegation, the original contracting party remains bound, albeit that performance of the contractual obligations is carried out by another party.

As with assignation, the existence or otherwise of *delectus personae* will determine whether or not the obligation of performance can be delegated. The work content of the contract will assist in determining whether or not delegation is competent. Work which consists chiefly of manual labour has been held to contain no element of *delectus personae*, see *Asphaltic Limestone Concrete Co Ltd and Another* v. *Corporation of the City of Glasgow* (1907). However, it has been recognised that the execution of repair work on and in residential properties is of a character which might well be the subject of *delectus personae*, see *Scottish Homes* v. *Inverclyde DC* (1997).

JCT 98 contemplates delegation. Clause 19.2.2 permits the sub-letting of any portion of the contract works. That, however, cannot take place without the written consent of the architect. Such consent is not to be unreasonably delayed or withheld. The underlying principle of responsibility resting with the original contracting party is echoed in clause 19.2.2 which specifies that the main contractor shall remain wholly responsible for the carrying out and completing the works in accordance with the contract, notwithstanding the sub-letting of any portion of the contract works.

## 11.7 Novation

Novation is, broadly, the substitution of a new party for an existing party to a contract, with the result that the new party assumes the rights and obligations under the contract. Novation requires the consent of each of the parties to the original contract. That consent need not be express and can be inferred from the conduct of the parties. See *McIntosh & Son* v. *Ainslie* (1872). Strictly speaking, novation has the effect of simultaneously extinguishing the old obligation and creating a new obligation. The effect of this must be carefully considered by the parties before proceeding with novation. For example, claims for breach of the original contract may be extinguished by novation, see *Hawthorns & Co Ltd* v. *Whimster & Company* (1917).

Novation may be of particular relevance where the contractor becomes insolvent. One of the options open to the employer, and the insolvency practitioner responsible for the affairs of the contractor, is the novation of the original contract to a substitute contractor, whether on the same terms as the existing contract or on varied terms. The Insolvency Practice Guide produced by SBCC provides assistance to employers and their professional advisers upon the event of the insolvency of a main contractor whilst carrying out work under a Scottish Building Contract. This includes guidance upon novation and a suggested style of novation agreement. In construction insolvency, a true novation, that is where the substitute contractor steps into the position of the original contractor and the contract continues as if the substitute contractor had been the original contractor, is most unlikely. Considerations such as liability for defects and the potential liability for liquidated and ascertained damages will militate against true novation. In most cases, therefore, the transaction will be a conditional novation, in terms of which the substitute contractor takes on only certain limited obligations, e.g. the obligation to complete, and does not assume responsibility for antecedent breaches.

# Chapter 12
# Collateral Warranties

## 12.1 *Introduction*

The use of collateral warranties in the construction industry has increased greatly in recent years. In this chapter it is intended to provide some practical comments on collateral warranties, firstly by looking at the respective interests of the various parties involved in a typical building project which have generated the need for collateral warranties, and secondly by considering certain of the clauses which are frequently to be found in collateral warranties.

The decision of the House of Lords in *Murphy* v. *Brentwood DC* (1991) could not be said to have been the catalyst in the creation of the collateral warranty 'industry', as the practice of requiring collateral warranties almost as a matter of course in construction developments had evolved a number of years earlier. However, *Murphy* reinforced the need for such a practice.

For a further discussion of *Murphy* and related cases, see Section 9.10.

A practice had arisen in the early 1980s of requiring contractors, subcontractors and construction professionals to acknowledge, by contractual means, duties of care to parties with whom they had otherwise no contractual relationship, for example tenants, funders and subsequent owners. The perceived inadequacies and uncertainties of a remedy based upon delict led to the practice of creating, by means of a collateral warranty, a contractual nexus, which would otherwise be absent. The purpose of a collateral warranty is to impose by contract duties and obligations on the part of the contractor, sub-contractor or consultant in favour of a third party who is not the original building owner or employer but who may nevertheless suffer loss in the event of a construction or design defect.

Collateral warranties essentially deal with the apportionment of risk. Negotiation of the terms of collateral warranties has become a mini-industry in its own right, with the need to satisfy the conflicting interests of the beneficiary under the collateral warranty (referred to in this chapter as the 'warrantee') and the grantor (referred to as the 'warrantor'). Bodies within the construction industry, including the Scottish Building Contract Committee, have made various attempts to satisfy these conflicting interests without the need for protracted negotiation by producing standardised forms of collateral warranty.

As yet there are few reported cases involving collateral warranties, see *Hill Samuel Bank Ltd* v. *Frederick Brand Partnership* (1994). It remains to be seen whether there will be an adverse reaction from the insurance market to the continued use of collateral warranties in the event of a judicial decision on collateral warranties which is unfavourable to insurers' interests.

## 12.2 *Interests in obtaining warranties*

The reasons why parties involved in a construction project require collateral warranties vary depending upon their particular position in the project.

### 12.2.1 The developer

In a typical commercial development the developer of the project will intend either to realise his investment at, or shortly after, completion of the project by disposal to a third party purchaser, or to grant a leasehold interest to one or more tenants. In either case the marketability of the development will demand that collateral warranties from the contractor and the consultants are available to the purchaser and/or tenants.

### 12.2.2 The funder

Whilst the funder will normally be protected by a heritable security over the development, it will wish to preserve a right of recourse against any party whose actions may diminish the value of that security. In addition, the funder will wish to have the option of ensuring that the development is completed (and the value of the security therefore maximised) in the event of the developer becoming insolvent. This is achieved by exercising 'step-in rights', (see Section 12.3.6 below).

From a funder's perspective, a collateral warranty from the quantity surveyor is also important. Interim valuations under the building contract will normally be carried out by the quantity surveyor on behalf of the developer and will, ordinarily, be reflected in payments to the contractor. In turn drawdowns from the funder's loan to the developer will typically reflect the amount of such valuations and payments, hence the need for the funder to secure some degree of comfort in relation to the actings of the quantity surveyor.

### 12.2.3 Purchaser/tenant

Since the decision in *Murphy* it is clear that, except in very specialised circumstances, a purchaser or tenant will have no claim in delict against a contractor or consultant with whom the purchaser or tenant has no contract. Hence the need for purchasers and tenants to protect themselves against the manifestation of latent defects caused by faulty design or construction. The principle of *caveat emptor* (buyer beware) applying in contracts for the sale of heritable property prohibits any recourse against a seller, and the obligations undertaken by tenants under a typical commercial full repairing lease will require them to make good all damage to the leased property howsoever caused (notwithstanding that it may be due to a latent defect) without recourse to the landlord.

### 12.2.4 Drawbacks of assignation

It is not unusual for developers to offer to assign to a purchaser or tenant their rights against the contractor and/or consultants responsible for constructing and designing the development (and express provision is made for this in clause 19.12 of JCT 98). However assignation is very much a poor and unsatisfactory substitute for a collateral warranty. This is because the principle *assignatus utitur jure auctoris* applies to assignation, i.e. the assignee stands in the shoes of the assignor, can acquire no better rights than the assignor, and is subject to any defence available against the assignor. Thus, if the developers have sold their interest to a purchaser for full value before any defect in the development became apparent, they cannot truly be said to have suffered any loss, the right of recovery of which is capable of being assigned to the purchaser. Likewise it is difficult to see how developers, who have let their property under a full repairing lease and who can demand rectification of a latent defect by the tenant, have suffered loss in the event of such a latent defect arising.

### 12.2.5 Nominated sub-contracts

Collateral warranties from domestic sub-contractors in favour of interested third parties may be regarded as 'belt and braces'. In other words primary liability will lie with the main contractors under their collateral warranty and additional warranties from domestic sub-contractors will need to be enforced only in the event that recovery cannot be made from the main contractor, notwithstanding that liability is established.

However, there is an important difference in the case of nominated sub-contractors. Under JCT 98 a main contractor has no direct liability to the employer for any design carried out by a nominated sub-contractor and the employer must ensure that it enters into the Nominated Sub-Contract Agreement with the sub-contractor to preserve a direct right of recourse for a design failure. It therefore follows that main contractors (since their liability under the collateral warranty is predicated on their liability under the main contract) will have no liability to the relevant warrantee for the design of a nominated sub-contractor. To protect themselves against a failure of such design, warrantees should therefore obtain collateral warranties from the relevant nominated sub-contractors.

### 12.2.6 Design and build contracts

In a design and build contract the issues described above at Section 12.2.5 in relation to nominated sub-contractors will not apply. However, it is nonetheless common practice for external professional consultants engaged by the design and build contractor to be required to give collateral warranties to the usual third party warrantees (i.e purchaser, funder and tenants), in addition to the warranties from the contractor to those parties, and also to the employer. This again may be regarded as a belt and braces approach, but with perhaps some justification in this instance, given the risk of the contractor becoming insolvent and the availability of professional indemnity insurance (in most cases) to back up claims made under such warranties.

## 12.3 Typical clauses

The following are examples of typical clauses found in collateral warranties. However, these are by no means exhaustive and the drafting of provisions intended to have the same effect may differ considerably from one warranty to another.

### 12.3.1 Standard of care

> 'The [Consultant/Contractor/Sub-Contractor] warrants that it has exercised and will continue to exercise the reasonable skill, care and diligence to be expected of a suitably qualified and competent [Consultant/Contractor/Sub-Contractor] experienced in projects of a similar size, scope and complexity to the Development in the performance of its services to the Employer under the Appointment/Building Contract/Sub-Contract ("the Primary Contract").'

From the point of view of the warrantor, it is important that the standard against which its responsibilities will be measured for the purposes of the collateral warranty is no higher than the standard of the Primary Contract. A warrantor often wishes to restrict, either in quantum or nature, the losses for which it may be liable under the collateral warranty. For example, it is common, particularly in the standard forms of warranties (and is often demanded by professional indemnity insurers) that the warrantee's liability be restricted to the reasonable cost of repair, renewal or reinstatement of the development, to the extent attributable to the warrantor's breach. Any liability, for example, for loss of use would therefore be excluded.

### 12.3.2 Deleterious materials

> 'The [Consultant/Contractor] shall use reasonable skill and care to ensure that no substance generally known at the time of specification/use to be

> deleterious to health or safety or to the durability of the works or not in accordance with the British or equivalent standards or codes of practice shall be specified for or used in the construction of the Development.'

Until recently the practice for many years had been to specify a list of materials not to be used, or specified for use, in the development. These lists were often prepared with little thought and lawyers were criticised (often with justification) for drafting collateral warranties containing lists of prohibited materials without any regard for the nature of the project or the particular properties of a prohibited material. A blanket prohibition may not be appropriate for a particular material, depending on the nature of the project and the intended use of the material. The matter was brought to a head when a manufacturer of calcium cilicate bricks successfully obtained an interim interdict against a local authority preventing it from including calcium cilicate bricks in the list of deleterious materials in its collateral warranties. It is now common practice to use a more general clause.

Parties should be clear as to whether the warrantor is warranting that the materials have not been used during the execution of the development, or whether it is warranting only that it has used reasonable skill, care and diligence to ensure that they have not been used, or have not been specified for use as the case may be. It should also be borne in mind that references to 'good building practice' are potentially subjective and, in terms of scope, very wide.

Knowledge of the deleterious nature of the material is usually stated to be tested at the time of either specification or use of the material in question. There is some danger in a consultant or contractor accepting the time of use as the relevant benchmark, given that the material may have been capable of being used legitimately or its deleterious properties may not have been known at the time of specification, and yet the consultant/contractor will be in breach of the warranty if that state of affairs changes prior to the use of the material in the development.

### 12.3.3 Limitations on proceedings

> 'No action or proceedings for any breach of this Agreement shall be commenced against the [Consultant/Contractor] after the expiry of [ ] years from the date of practical completion of the Development as certified under the Building Contract.'

A clause of this type is particularly important where the warrantor does not enter into the collateral warranty until after practical completion. Failure to include such a clause could potentially leave the warrantor owing contractual duties to a third party of longer duration than those owed to the client under the Primary Contract as, in the absence of such a provision, the prescriptive period may not then commence until the date of execution of the collateral warranty, notwithstanding that the design or construction failure giving rise to breach of the warranty occurred prior to the execution of the warranty.

It is sometimes otherwise difficult to see a rational justification for imposing a limitation period in a collateral warranty which reduces the warrantor's period of exposure to claims (at least in the case of latent defects) to a period usually significantly less than the statutory prescriptive period (see Section 8.9). Ultimately it is a commercial matter. The warrantor will argue that, as the warrantee is being granted rights beyond those otherwise available to it at law, there is sound commercial sense for the warrantor to place limitations on the period during which such rights may be exercised. Pragmatic considerations may also apply. For instance, a funder may consider it unnecessary to require a limitation period longer than the period for repayment of the loan.

### 12.3.4 Net contribution

> 'The liability of the [Consultant/Contractor] for costs under this Agreement shall be limited to that proportion of the Company's losses which it would be just and equitable to require the [Consultant/Contractor] to pay having regard to the extent of the [Consultant/Contractor's] responsibility for the same and on the basis that [list names of other Consultants/ Contractor/ Sub-Contractor] shall be deemed to have provided contractual undertakings on terms no less onerous than this Clause [ ] to the Company in respect of the performance of their services in connection with the Development and shall be deemed to have paid to the Company such proportion which it would be just and equitable for them to pay having regard to the extent of their responsibility.'

The above is commonly known as a 'net contribution' clause and is intended to alleviate what are regarded (at least in the eyes of warrantors and their insurers) as the harsh consequences of joint and several liability. This arises where damage to the warrantee is caused as a result of breach of duty of more than one party. If each breach of duty has materially contributed to the same damage to the warrantee, the warrantee is entitled to recover the losses arising from such damage from any or all of the parties in breach. Thus one consultant may (in the absence of a net contribution clause) be pursued for the whole of the loss, notwithstanding that other consultants or contractors may have contributed to that loss.

The clause creates the fiction that all the relevant parties are deemed to have granted collateral warranties (whether, in reality, they have or not) to the warrantee. To the extent that they have a responsibility, they are deemed to have already paid their fair share of the recoverable loss or damage suffered by the warrantee. The relevant warrantor is then left liable only for its share of the loss on the basis that the other warrantors are deemed to have paid their contribution.

This clause is intended to reflect the principles of s.3 of the Law Reform

(Miscellaneous Provisions) (Scotland) Act 1940 which gives the court the power to apportion damages against joint wrongdoers, and the clause may, on the face of it, seem equitable. However, unlike the Act, the clause places on the warrantee the onus of obtaining similar warranties from the other parties 'on terms no less onerous' and also the risk that recovery may not in fact be made from one of the relevant parties, notwithstanding that liability of the other parties is calculated on the basis that it is deemed to be so made.

The clause remains untested by the courts.

There have been some concerns expressed regarding evidence that may be admitted by the courts and the potential of a decree being granted against an unrepresented party. Some professional indemnity insurers will insist upon the inclusion of a net contribution clause in collateral warranties before extending cover to liabilities arising out of them.

### 12.3.5 Equivalent rights of defence

> 'The [Consultant/Contractor/Sub-Contractor] shall be entitled in any action or proceedings by the [Company] to rely on any limitation in the Appointment and to raise the equivalent rights in defence of liability as it would have against the Employer under the Appointment, had the [Company] been named as Employer under the Appointment.'

The intention of this clause is to ensure that liability under the collateral warranty is co-extensive with that under the Primary Contract. However, in the drafting of this type of clause (which may be worded in a number of different ways) care must be taken to ensure that it does not confer a 'no loss' right of defence on the warrantor. For example, under the above wording, is there perhaps an argument that if the warrantee (being, say, a subsequent purchaser) had been named as original employer the warrantor would be entitled to contend it had no liability because at the time of the appointment being entered into, the employer had no proprietary interest in the development?

### 12.3.6 Step-in rights

The rights contained in this clause allow the warrantee to step into the shoes of the employer in certain pre-defined situations, typically insolvency events or where the warrantor seeks to exercise its right to determine the principal contract.

Step-in provisions will normally appear only in a funder's collateral warranty, although in some situations they may also be appropriate in a purchaser's collateral warranty. Step-in allows the warrantee to step into the shoes of the employer under the Primary Contract by way of novation. Novation is considered above in Section 11.7. It is generally activated where the warrantee wishes to prevent the warrantor from terminating the Primary

Contract or otherwise wishes to ensure completion of the project following default by the developer.

The warrantee will normally be obliged to assume all the outstanding and future obligations of the employer, for example, payment of outstanding and future fees. The warrantee has the right but not the obligation to step-in and therefore there will generally be a specified time within which it may receive all of the relevant information to allow it to reach an informed view as to whether it wishes to exercise its right or not. Warrantors will wish to keep this period to a minimum as they are otherwise prevented from terminating during this period, even although they may not be getting paid.

The warrantor will usually require an assurance that the employer protects it from any action where the warrantee invokes its step-in rights. For this reason, collateral warranties containing step-in provisions will normally be tripartite, with the employer, warrantee and warrantor being signatories.

### 12.3.7 Assignation/obligation to enter into further warranties

A warrantee will normally require that a collateral warranty in its favour is capable of being assigned. In the ordinary course of events, the warrantor, to limit its exposure, will attempt to limit the number of occasions upon which the collateral warranty may be assigned. Both warrantor and warrantee should check the provisions of the warrantor's professional indemnity cover when considering the number of assignations as most insurers will wish to restrict the permitted number of assignations, especially where the collateral warranty contains few restrictions on liability.

For the reasons described in section 12.2.4 above, an assignation of a collateral warranty is of limited value and a second purchaser or tenant would normally prefer a new warranty in its favour.

### 12.3.8 Insurance

> 'The Consultant shall use reasonable endeavours to maintain professional indemnity insurance in an amount of not less than £ [ ] for every occurrence or series of occurrences arising out of any one event with insurers of substance and repute in the UK Insurance Market until [ ] years from the date of Practical Completion under the Building Contract, provided always that such insurance is available to the Consultant at commercially reasonable rates. The Consultant shall immediately inform [*x*] if such insurance ceases to be available at commercially reasonable rates in order that the Consultant and [*x*] can discuss the means of best protecting the respective positions of [*x*] and the Consultant in the absence of such insurance. As and when it is reasonably requested to do so by [*x*, or its appointee], the Consultant shall produce for inspection documentary evidence that its professional indemnity insurance is being maintained.'

The well advised warrantor will refer any collateral warranty it proposes to enter into to its professional indemnity insurers for comment. Typically, the warrantor will be required to obtain professional indemnity cover and maintain it for a number of years after practical completion. Further qualifications to this obligation include the requirement for insurance to be available at commercially reasonable rates and on reasonable terms and conditions.

It is worth noting that there are few, if any, insurers prepared to underwrite any form of absolute risk. Professional indemnity insurance traditionally covers the 'legal liability' of the insured, i.e. in the case of a professional, the exercise of reasonable skill and care. Any voluntary assumption of a greater duty by the insured in contract is generally covered by way of an extension to the policy. This can either be a general endorsement within certain parameters, or require 'approval' of each contract by the insurer.

### 12.3.9 Intellectual property licence

Most collateral warranties will contain a non-exclusive licence by the warrantor in favour of the warrantee in respect of the use of intellectual property rights in the design documentation prepared relative to the development. This may be particularly important, for example, to a purchaser who wishes to construct an extension to the property consistent with the existing design.

## 12.4 *Effects of the 1996 Act on collateral warranties*

Are collateral warranties agreements for the carrying out of construction operations in terms of s. 104(1)(a) of the 1996 Act?

Conflicting views have been expressed to date on the answer to this question, and of course if collateral warranties do indeed fall within the ambit of s. 104(1)(a), disputes under warranties would, in the absence of express adjudication provisions, fall to be determined by the statutory adjudication scheme. There is therefore a practical argument in favour of collateral warranties containing dispute resolution provisions which are compliant with the 1996 Act and also consistent with any specific adjudication provisions contained in the principal construction contract or consultants' appointment (as the case may be). Even leaving aside the question whether collateral warranties fall within the scope of the 1996 Act, there are practical advantages in having dispute resolution procedures in a collateral warranty which are consistent with those in the principal contract (although it has to be said that prior to the 1996 Act there did not appear to be any real attempt in the standard forms of collateral warranty to achieve this).

The Scottish Building Contract Committee has published revisions to its standard forms of collateral warranty by main contractor and sub-contractor which insert 1996 Act compliant dispute disposal provisions.

# Chapter 13
# Insurance, Guarantees and Bonds

## 13.1 *Insurance*

### 13.1.1 Insurance under building contracts

Although this chapter concentrates on the insurance provisions contained in JCT 98, most building contracts contain insurance provisions which are broadly similar. In the majority of building contracts the contractor undertakes to indemnify the employer for loss and liabilities arising from death of or injury to persons and loss of or damage to property, and the contractor will be obliged to maintain employer's liability and public liability insurance to cover the risk of such loss or liability occurring.

Such types of insurance fall within the category of 'liability' insurance. In other words the insurance will cover liability which the insured party has to a third party as a result of the insured event.

Most construction contracts will also expressly deal with the other common category of insurance, namely property insurance. In the context of construction contracts this type of insurance will cover the contract works, site materials, plant and equipment. The obligation to take out and maintain such insurance may be dealt with in differing ways. Thus, while in most cases insurance will need to be in the joint names of employer and contractor, some contracts may provide that the obligation to take out such insurance is that of the employer, while others may impose that obligation on the contractor.

It should also be borne in mind that the characteristics of certain types of contract may demand more extensive insurance requirements. For example, insurance may need to be taken out for certain contracts (but not others) relating to business interruption, fortuitous pollution, marine claims and/or professional indemnity. At the same time not all risks are insurable (or at least not under conventional policies). An obvious example of this is a construction defect not involving a design error. While this may be insurable under a specialist latent defects policy (section 13.1.16), it will not be covered by a standard property insurance policy or a professional indemnity policy.

### 13.1.2 Definition of insurance

Broadly speaking, a contract of insurance is a contract whereby, for a consideration (necessary at least under English law and normally payment of premiums), the insured obtains a benefit (usually payment of money) upon the happening of a certain event in respect of which there is uncertainty as to either whether it will happen or when it will happen. Finally, the insurance must be 'against something'. See *Prudential Insurance Co* v. *IRC* (1904).

### 13.1.3 Legal characteristics of insurance

The requirement that insurance must be 'against something' is generally taken to mean that the insured must have an 'insurable interest'. This means that the insured must have a pecuniary interest in the subject matter of the insurance so that upon the occurrence of the insured event the insured has, as a result, either himself suffered a loss or incurred a legal liability.

The other fundamental principle to which all insurance contracts are subject is that of *uberrimae fidei*, or utmost good faith. This principle requires each party to make a full disclosure of all material facts which may influence the other party in deciding to enter into the contract. A failure to disclose such material facts may render the policy void, which in practical terms would allow the insurer to refuse to meet a claim. This would apply even in the absence of fraudulent intent.

The principle of subrogation is common to all insurance contracts which involve the insurer indemnifying the insured in respect of a loss or a liability. Subrogation means that the insurer is entitled to exercise any remedy which may have been exercisable by the insured in respect of the insured event. In practice, it means that the insurer can pursue a claim (in the name of the insured) against a third party who may be responsible, either wholly or partly, for the insured loss. Such a right is subject to the insurer having made payment in respect of the insured's claim and to subrogation rights not having been excluded by any express contractual term. It should also be noted that subrogation rights are not available against a party who is a joint insured under a joint names insurance policy, see *Petrofina (UK) Ltd and Others* v. *Magnaload Ltd and Others* (1983). An example of this is the joint names policy to be effected under clause 22 of JCT 98, considered below in section 13.1.8.

### 13.1.4 Indemnities and insurance

It is important to recognise the distinction between indemnity and insurance. A building contract will normally contain provisions in terms of which the contractor will undertake to indemnify the employer against the occurrence of certain risks. The contract will also impose obligations on either or both of the parties relative to insurance of risks. There is a crossover between these indemnity and insurance obligations, insofar as insurance may be required to be taken out against risks covered by certain indemnities, but the obligations are not necessarily co-extensive. Risks covered by a particular indemnity may not necessarily be insurable (and this is a matter which it is prudent for the party undertaking the indemnity to check).

Further, the parties may agree that a certain risk be covered by insurance and that neither party should have liability notwithstanding fault.

Clause 20 of JCT 98 broadly imposes an obligation on the contractor to indemnify the employer against

(i) death and personal injury arising out of the carrying out of the works, except to the extent that the death or injury is due to the act or neglect of the employer or any person for whom he is responsible; and
(ii) property damage arising out of the carrying out of the works and to the extent that the damage is due to any default of the contractor or other persons properly engaged on the works or on site in connection with the works (other than the employer and persons engaged or authorised by him or local authorities and statutory undertakers executing work in pursuance of their statutory rights and obligations).

### 13.1.5 Insurance under JCT 98

JCT 98 imposes (or, in some cases, gives an option to impose) obligations to take out and maintain insurance covering the following five types of risk:

- personal injury and death (see Section 13.1.6)
- damage to property (other than the works) arising from the contractor's default (see Section 13.1.6)
- 'non-negligent' damage to property (other than the works) (see Section 13.1.7)
- damage to the works (see Section 13.1.9)
- loss of liquidated and ascertained damages (see Section 13.1.13).

'Excepted Risks' are excluded from the obligation to insure (see Section 13.1.14).

There are also other types of insurance cover not contemplated by JCT 98, e.g. professional indemnity insurance and latent defects insurance (see Sections 13.1.15 and 13.1.16) and a well-advised party to a construction contract will consider whether any such (and other) risks should be covered by insurance, notwithstanding the fact that there may be no contractual obligation to do so.

### 13.1.6 Insurance against injuries to persons or damage to property

Under clause 21.1 of JCT 98, the obligation to insure against the death of, or injury to, any person or loss of, or damage to, any property arising out of or in consequence of the execution of the works is imposed on the contractor. This reflects the express indemnity given by the contractor under clause 20.1 for such injury damage or loss (see Section 13.1.4). The cover is usually contained in two separate policies, namely a public liability policy and an employer's liability policy.

Insurance relating to personal injury or death of an employee of the contractor must comply with the Employers' Liability (Compulsory Insurance) Act 1969 which specifies a statutory minimum cover.

Insurance cover in respect of the death of or injury to other persons and loss of or damage to property will be effected under a public liability policy. The minimum amount of public liability cover in respect of any one occurrence should be stated in the appendix to the building contract.

The insurance must remain in force until practical completion.

### 13.1.7 Clause 21.2.1 insurance

There is a further option open to the employer under clause 21.2.1 of JCT 98. This insurance need only be taken out by the contractor if the appendix to the contract so specifies.

This insurance relates to damage to property, other than the works, and to site materials, caused by certain specified risks, namely:

- collapse,
- subsidence,
- heave,
- vibration,
- weakening or removal of support, or
- lowering of ground water

arising out of or in the course of carrying out the works.

This cover is most commonly required when there is neighbouring property susceptible to damage by any of the above risks.

Under clause 21.2.1 fault does not need to be established but there are a number of exceptions which reduce considerably the scope of this insurance cover, namely:

- Injury or damage for which the contractor is liable under clause 20.2 (which should be insured under clause 21.1).
- Injury or damage attributable to errors or omissions in the designing of the works.
- Injury or damage which can reasonably be foreseen to be inevitable, having regard to the nature of the work to be executed or the manner of its execution.
- Injury or damage which is the responsibility of the employer to insure under clause 22C.1 (insurance of existing structures).
- Injury or damage arising from war risks or the 'Excepted Risks' (see Section 13.1.14).

### 13.1.8 Insurance of the works

JCT 98 provides three options for insuring the works. A choice must be made in the appendix to the contract. However, certain provisions apply regardless of which option is selected.

The contract calls for a 'joint names policy' for 'all risks insurance'. Where the policy is in joint names both the contractor and the employer are named as an insured and either may make a claim under the policy in its own name. The insurer has no right of subrogation against either party (see Section 13.1.3).

The all risks insurance which the contractor or employer, as the case may be, is obliged to take out should provide cover against any physical loss or damage to work executed and site materials, but *excluding* the costs necessary to repair, replace or rectify:

(a) property which is defective due to wear and tear, obsolescence, deterioration, rust or mildew;
(b) any work executed or any site materials lost or damaged as a result of its own defect in design, etc.;
(c) loss or damage caused by or arising from the consequences of war, invasion, rebellion, nationalisation, etc., disappearance or shortage (if such disappearance or shortage is only revealed when an inventory is made or is not traceable to an identifiable event), or an 'excepted risk'.

There are certain further exclusions applicable to Northern Ireland only.

### 13.1.9 All risks insurance by the contractor

If clause 22A of JCT 98 is selected, the contractor must take out and maintain a joint names policy for all risks insurance for the full reinstatement value of the works plus a percentage, if any, to cover professional fees as stated in the appendix to the contract. This joint names policy must be maintained up to the date of issue of the certificate of practical completion or the date of determination of the employment of the contractor, whichever is the earlier.

An alternative open to the contractor, and which is widely used by many contracting companies, is to use an existing annual policy which complies with the obligations in clause 22. However, the policy must still be a joint names policy and the contractor must provide documentary evidence that the policy is being maintained and, when so required, supply for inspection the policy itself and the premium receipts.

### 13.1.10 All risks insurance by the employer

The second option, contained in clause 22B, is for the employer to take out the joint names policy for all risks insurance on the same terms and for the same period as described above. There is a corresponding provision to that contained in clause 22A, entitling the contractor to take out the joint names policy if the employer fails to do so.

### 13.1.11 Existing structures

The third option, in clause 22C, applies where the contract is for alteration or extension to existing structures owned by the employer or for which he is responsible. In this case the employer takes out and maintains the joint names policy for all risks insurance for the works themselves (as per clause 22B) but must also maintain a joint names policy for **existing structures** to cover the cost of reinstatement, repair or replacement of loss or damage due to one or more of the 'Specified Perils'. If the employer fails to take out either of these two insurances the contractor is entitled to do so and recover the cost of the premiums. The 'Specified Perils' are defined as fire, lightning, explosion, storm, tempest, flood, bursting or overflowing of water tanks, apparatus or pipes, earthquake, aircraft and other aerial devices or articles dropped therefrom, riot and civil commotion, but excluding 'excepted risks'.

Clause 22C, and its inter-action with the contractor's obligations to indemnify the employer against property damage to the extent that it is due to default of the contractor, has produced some interesting results. The equivalent to clause 22C.1 under JCT 1963 stated that existing structures and the contents thereof were at the 'sole risk' of the employer. In the Scottish appeal of *Scottish Special Housing Association* v. *Wimpey Construction UK Ltd* (1986) the House of Lords held that the effect of this wording was that the employer was bound to insure the property against the risk of damage by all of the Specified Perils, including fire, and that the liability for such damage rested with the employer, notwithstanding that the fire was caused by the negligence of the contractor.

JCT 98 does not expressly state that the existing structures are to be at the employer's 'sole risk' as regards the Specified Perils. However, the contractor's obligation under clause 20.2 to indemnify the employer against damage to property is still stated to be 'subject to ..., where applicable, clause 22C.1'. This is thought to mean that the employer continues to bear the whole risk of damage to the existing structures caused by the 'specified perils' notwithstanding the contractor's negligence; otherwise the words 'subject to ... clause 22C.1' would have no meaning.

This was considered in passing in the recent House of Lords decision in the Scottish appeal of *British Telecommunications plc* v. *James Thomson & Sons (Engineers) Ltd* (1999) where initially in a lower court counsel for BT had conceded that, had the damage to the existing structure been caused by the main contractor, BT (the employer) would have assumed the risk (of the Specified Peril) at the outset. Counsel for BT did not repeat this concession in the House of Lords as a result of another recent case, *Kruger Tissue (Industries) Ltd* v. *Frank Galliers Ltd* (1998), where it was held that a negligent contractor could be liable for consequential losses suffered by an employer notwithstanding that the damage occurred to an existing structure covered by a clause 22C.1 joint names policy. It is an open issue, therefore, whether *Scottish Special Housing Association* would be followed under the JCT 98 wording.

While it may not be clear whether the employer has a right of indemnity against a negligent contractor where insurance has been effected under clause 22C, it does appear clear that the employer has no obligation to indemnify the contractor against claims by third parties against the contractor, in respect of this same damage. See *Aberdeen Harbour Board* v. *Heating Enterprises (Aberdeen) Ltd* (1990).

### 13.1.12 Claims

The consequences of loss of or damage to the works under JCT 98 vary according to which insurance option has been selected.

If the contractor has taken out insurance under clause 22A the contractor must give notice of the loss or damage to the architect or to the employer, and, following any inspection by the insurers, the contractor must restore the damaged work and repair or replace any lost or damaged site materials. The contractor must authorise the insurers to pay any monies to the employer, who then passes them on to the contractor by way of interim certificates. No other sums are payable to the contractor by virtue of the loss or damage (this would appear to exclude a claim for loss and expense under clause 26), although an extension of time may be granted if the loss or damage was occasioned by any of the Specified Perils (clause 25.4.3.).

Where the all risks insurance has been taken out by the employer under clause 22B there is a similar procedure for the contractor to give notice, but any restoration, replacement or repair work is to be treated as a variation under clause 13.2 and valued accordingly. In this case any shortfall between insurance proceeds and the cost will be made up by the employer. Under clause 22A, in contrast, the contractor will only receive such monies as the insurers pay out.

Where loss or damage affects work being carried out to existing structures which have been insured by the employer, there is an option under JCT 98 for either party to determine the employment of the contractor 'if it is just and equitable so to do'. Any dispute on this point shall be referred for resolution by adjudication, arbitration or the court. If no notice of determination is served, or where the adjudicator, arbiter or court decides that it was not just and equitable to terminate the contractor's employment, then the contractor must restore, replace or repair the loss or damage, such work being treated as a variation under clause 13.2. There is little authority on what is meant by 'just and equitable' but it would appear to include commercial viability of the project from either party's perspective.

### 13.1.13 Insurance for employer's loss of liquidated and ascertained damages

Under clause 22D of JCT 98, the employer has the option to require the contractor to take out a policy which will compensate the employer for inability to recover liquidated and ascertained damages for a period of delay for which any extension of time has been granted under clause 25.4.3, namely loss or damage occasioned by any one or more of the 'Specified Perils'. For example, if a fire at the site has delayed the contractor in completing the works then he would ordinarily be entitled to an extension of time on this ground and the employer would not be entitled to levy liquidated and ascertained damages.

The appendix to the contract will state whether or not such insurance **may** be required. As soon as the contract has been entered into the architect will inform the contractor either that no insurance is required or instruct him to obtain a quotation at an agreed value. The contractor then obtains the quotation which is forwarded to the architect. The architect seeks instructions from the employer and then informs the contractor whether or not the quotation should be accepted. Once the policy is issued it must be deposited with the employer together with a receipt for the premium. The cost of the premiums is added to the contract sum.

### 13.1.14 Excepted risks

All of the provisions in clause 20.1, 20.2 and 21.1 of JCT 98 must be read subject to clause 21.3 which excludes from the contractor's liability, death, injury, loss or damage caused by the effect of an excepted risk. These are defined in JCT 98 as ionising radiations or contamination by radioactivity from any nuclear fuel or from any nuclear waste from the combustion of nuclear fuel, radioactive, toxic, explosive or other hazardous properties of any explosive nuclear assembly or nuclear component thereof, and pressure waves caused by aircraft or other aerial devices travelling at sonic or supersonic speeds.

### 13.1.15 Professional indemnity insurance

Discussion on the terms and conditions of appointment of construction professionals is outside the scope of this book. However, with the continuing demand for design and build packages, professional indemnity or design liability insurance for contractors has become an integral part of the building contract structure in terms of spreading risk.

Under a design and build contract the contractor assumes responsibility for both design and construction of the project and, in the absence of in-house designers, will sub-contract design to one or more professional firms of architects, structural engineers, services engineers, etc. However, the employer's contract remains with the building contractor alone - although a prudent employer would also seek collateral warranties from each member of the design team. In the event of a defect in design the employer's primary claim will lie against the contractor and it is for this reason that the contractor would be well advised (if not obliged under the contract) to maintain professional indemnity insurance to cover any such claim. The contractor or his insurers may have a right of recovery against the design consultants and/or their professional indemnity insurers.

Perhaps somewhat curiously, neither JCT nor most of the other publishers of standard forms of design and build contracts have included in their standard forms obligations on the contractor to maintain professional indemnity insurance (although such an obligation is common in collateral warranties), and to date there is no requirement for such insurance in the JCT Standard Form of Contract with Contractor's Design 1998 edition or the

Scottish equivalent, the Scottish Building Contract With Contractor's Design, or in the Contractor's Designed Portion Supplement to the Standard Form of Contract.

Notwithstanding this, most medium to large firms of contractors who are engaged in any kind of design work carry professional indemnity insurance.

It is worthwhile noting the extent of the contractor's liability under the Scottish Building Contract with Contractor's Design. Clause 2.5.1 states that the contractor shall have 'the like liability to the employer ... as would an architect or as the case may be other appropriate professional designer holding himself out as competent to take on work for such design'.

This design warranty therefore equates the duty of the contractor to that of a professional designer, namely the duty to exercise reasonable skill and care. This removes any term, which may otherwise be implied, that the design is fit for the purpose for which it is required. Most insurers are not prepared to cover such a fitness for purpose warranty given by a design professional, and similarly are not prepared to cover such a warranty from a building contractor with design responsibility.

Professional indemnity policies are usually issued on a 'claims made' basis. This means that any claim will be dealt with under the policy in force during the year in which the claim is made. For that reason it is prudent for the contractor to maintain professional indemnity insurance long after completion of a project.

The amount of professional indemnity insurance cover should ideally be sufficient to cover any loss which is likely to result from a defect in design. Parties should, however, be aware that this is a continuing annual obligation. Notwithstanding the usual qualifications relating to availability of cover and the terms of renewal, parties should take care when agreeing the level of cover, bearing in mind the potentially volatile nature of the insurance market.

It is of course possible to limit the potential liability of the building contractor and indeed clause 2.5.3 of the Scottish Building Contract with Contractor's Design does permit the parties to insert a ceiling of liability for loss of use, loss of profit or other consequential loss.

### 13.1.16 Latent defects insurance

There is a separate type of policy which the employer himself can take out to cover physical damage to the works regardless of who is legally at fault. The policy will normally run from practical completion of the works for ten years (it is otherwise known as 'decennial insurance'), and would cover major physical damage or an inherent defect in the structure which threatens its stability. The defect must be inherent, i.e. existing from the outset but not discovered until after completion of the building.

As such a policy is intended to cover major damage, regardless of who is responsible, premiums tend to be high and this may explain why latent defects insurance is not currently widely used in the UK. Further, such policies do tend to contain exclusions which have not found favour in the UK. Also, from the contractor's perspective, unless such a policy contains a waiver of subrogation (available at additional cost) it has little benefit to him. Such insurance does, of course, have the advantage that, in the event of a design fault becoming apparent, it is not necessary for the employer to prove breach of contract and/or negligence on the part of the building contractor and/or design consultant. Although consequential loss extensions are available at additional cost, it should be noted that these policies will normally only cover physical damage to the building. Therefore a claim for negligent design, for example, on account of a building having a smaller net internal area than that set out in the employer's requirements, would have to be dealt with by a claim in the usual way against the building contractor or designer who produced the drawings.

## 13.2 *Guarantees and bonds*

### 13.2.1 Nature of bonds and guarantees

In broad terms a contract of guarantee is an undertaking by a person to secure the performance of the obligations of a party under a contract. A guarantee may (subject to the law of prescription) be unlimited as to time and amount. In construction contracts, guarantees of the contractor's obligations are typically provided by the contractor's parent company.

A bond is usually provided up to a maximum sum of money which will become payable in certain circumstances should one of the parties to a contract default. It is normally given by a bank or insurance company on behalf of either the employer - to cover payment to the contractor - or, more commonly, on behalf of the contractor to ensure payment of the damages arising in the event of the contractor's default. Normally, a performance bond will subsist only in respect of claims made prior to an end-date, typically either the date of practical completion or of making good defects. Thus, it will not normally, as a parent company guarantee often will, cover the cost of making good latent defects.

A bond may, except in certain circumstances where it is construed as an on-demand bond (see Section 13.2.3), be regarded as a form of guarantee, see *City of Glasgow DC* v. *Excess Insurance Company Ltd* (1986).

Despite the frequent use of guarantees in construction contracts, few of the institutions responsible for promoting standard forms of contract have published a form of guarantee (other than for performance bonds) and it is left to the parties to devise their own wording. In many cases it will be made a suspensive condition of the contract that such a guarantee be provided. Alternatively, the contract will be worded in such a way so as to make the failure to provide a guarantee a material breach entitling the other party to terminate the contract.

### 13.2.2 Guarantees - cautionary and principal obligations

A distinction may need to be drawn between a guarantee which is an independent obligation and one which is truly a 'cautionary obligation' and thus accessory to the principal obligation. The distinction can be important as, in the absence of express wording, variation of the principal contract may discharge a cautionary obligation. It is a question of fact whether the obligation is one of caution or a principal obligation, although clear wording should remove any doubt.

If the obligation is truly one of caution, it must be given the narrowest construction which the words will reasonably bear, see *Harmer* v. *Gibb* (1911). The cautioner's liability can never exceed that of the principal debtor and on payment of the debt a cautioner is entitled to recover from the principal debtor all sums which he has paid to the creditor. He is also entitled to demand from the creditor an assignation of the debt, any security held for it, and any diligence done upon it, so as to enable him to enforce his right of relief against the principal debtor.

Under s.6 of the Prescription and Limitation (Scotland) Act 1973 the prescriptive period applicable to cautionary obligations is five years.

The creditor should also have regard to any provision in the guarantee which provides for service of a demand on the guarantor.

Under s.8 of the Mercantile Law Amendment (Scotland) Act 1856, unless stated expressly to the contrary, there is no need for the creditor to pursue a remedy against the principal debtor before suing the cautioner (as was previously the position under common law).

In *De Montfort Insurance Co plc* v. *Lafferty* (1997), which related to a performance bond, it was held that a guarantor was not released from its obligations as a result of novation of the original building contract, the effect of which was to substitute the original employer with another party.

### 13.2.3 On demand bonds

Performance bonds fall into two general categories, namely on demand bonds and conditional bonds.

Unlike a guarantee, an on demand bond constitutes a primary obligation not dependent on first establishing the liability of a third party, i.e. (in most cases) the contractor. An on demand bond is payable upon the creditor's demand without any requirement to prove default or the amount of damages. The bondsman is then obliged to pay up to the level of the demand, subject to any monetary limit to the bond itself and subject to the demand complying with the terms of the bond. In the absence of fraud a court will not normally prevent enforcement of an on demand bond, see *Edward Owen Engineering Ltd* v. *Barclays Bank International Ltd* (1978). The party challenging the demand must be able to show that the only realistic inference from the facts is that the demand was fraudulent and that the bondsman was aware of the fraud. In most cases the bondsman will be prepared to pay upon demand, without challenge, because it, in turn, has obtained a counter indemnity from the contractor. It is therefore apparent that such bonds are open to abuse and are becoming less and less common, harder to agree, and very expensive (if available at all). It is fair to say they are used more in international contracts where a dispute over the conditionality of a bond (see below) is likely to be very difficult to resolve.

The question also arises as to the consequences to an improperly called bond. It is suggested that the employer should account to the contractor for the proceeds of the bond where the employer has not in fact suffered a loss in respect of the matter for which the bond was allegedly called.

In *The Royal Bank of Scotland Ltd* v. *Dinwoodie* (1987) the bondsman paid out monies pursuant to a performance bond. The bond was supported by a counter indemnity from individual guarantors who argued that the bank was wrong to make payment under the performance bond. On the wording of the bond, which guaranteed 'the damages sustained by the employer' by the contractor's default, it was held by the Outer House of the Court of Session that the surety bank was obliged to be satisfied, at the very least, that damages had been sustained by the employer and also as to the quantification of these damages.

### 13.2.4 Conditional bonds

A conditional bond will normally provide that it can only be called upon the occurrence of certain events, for example, the employer establishing that the contractor is in breach and the extent of the damages arising therefrom, or upon presentation of an arbiter's award or court decree or possibly the decision of an adjudicator pursuant to the 1996 Act.

There are a number of published standard forms of bond and two of these are considered below in more detail.

### 13.2.5 ABI Model Form Bond

This model form was published by the Association of British Insurers in September 1995, primarily in response to the English Court of Appeal decision in *Trafalgar House Construction (Regions) Ltd* v. *General Surety and Guarantee Co Ltd* (1994), in which the bond in question (similar to the ICE Bond considered below) was treated effectively as an on demand bond, despite earlier assumptions to the contrary. This decision was later overruled by the House of Lords, which followed *inter alia* the decision of *City of Glasgow DC* v. *Excess Insurance Co Ltd* (1986) that a performance bond in similar terms was a cautionary obligation.

The ABI Bond provides that the guarantor will 'satisfy and discharge the damages sustained by the employer as established and ascertained pursuant to and in accordance with the provisions of or by reference to the contract'. Thus the employer's entitlement is linked expressly to the contract itself. If, as with most standard forms, the contract contains a mechanism for ascertainment of loss and damages following breach by the contractor, then this mechanism must be followed before any money is payable under the bond. However, the wording of the bond does not go so far as to state that the employer has to establish the amount of his loss, if necessary by going to court or to arbitration, and it therefore leaves some uncertainty as to exactly at what stage and in what circumstances the bond can be called.

The ABI Bond does, however, expressly limit liability as to time and money. The 'expiry date' is a matter for negotiation but typically the date is stated as the date of practical completion of the works or the issue of the certificate of making good defects.

The clause also contains a prohibition on assignation by the employer without the prior written consent of the guarantor and the contractor. In practice this can lead to difficulties where, as in *De Montfort Insurance Co plc* v. *Lafferty* (1997), there is a change in the employer, either due to novation or because a funder has stepped into the building contract on the employer's default. This particular difficulty can be overcome by stating that the bond will be assignable to any successor to the employer under the contract.

The ABI Bond also makes clear that the bond operates as a guarantee, i.e. it is ancillary to the principal contract. Therefore if there is no liability or limited liability under the principal contract, liability under the bond will be similarly excluded or limited.

In order to avoid the potential difficulties associated with cautionary obligations, referred to above, it is made clear that the guarantor shall not be discharged or released by any alteration of any of the terms, conditions and provisions of the principal contract.

### 13.2.6 ICE Bond

An earlier form of bond was that published by the ICE and annexed to the ICE Conditions of Contract 5th and 6th Editions. This is similar to the type of bond described as 'archaic' by the Court of Appeal in *Trafalgar House Construction (Regions) Ltd* v. *General Surety and Guarantee Co Ltd* (1995). The operative part of this bond provides that 'on default by the contractor the surety shall satisfy and discharge the damages sustained by the employer thereby'. Although at first sight this is similar to that of the ABI Model Form the wording is not linked to the underlying contract and the *Trafalgar House Construction (Regions) Ltd* case brought into doubt the requirement for the employer to establish loss and damage by reference to the contract.

### 13.2.7 Time for claiming under a bond

The question also arises as to when the employer is entitled to make a claim under a conditional bond. This will of course depend upon the wording of the bond but in *Perar BV* v. *General Surety and Guarantee Co Ltd* (1994) the

English Court of Appeal held that insolvency and consequent termination of the contractor's employment was not a breach of contract which could trigger the bond. Although a termination event, insolvency was not itself a breach of the contract and the right to call for payment of the bond would not arise until, for example, the contractor had failed to make payment of any sums consequently due to the employer.

### 13.2.8 Retention bonds

Such bonds are becoming more commonplace as an alternative to the employer making a cash retention from the contract sum. The bond, backed up by a bank or insurance company, will secure the level of retention until the contractual date for release. If the contractor fails to honour its obligations to remedy defects the employer can call upon the bondsman to pay the requisite sum up to the maximum amount of the bond.

# Chapter 14
# Dispute Resolution

## 14.1 *Introduction*

Disputes frequently arise under building and engineering contracts. In this chapter, we will consider the available methods by which disputes can be resolved. In Scotland, the majority of construction industry disputes have, traditionally, been resolved by arbitration. That arises from the commonplace insertion of arbitration clauses into building and engineering contracts, and the right of a party to such a contract to insist upon arbitration in such circumstances. Nevertheless, the Scottish courts frequently become involved in building and engineering contract disputes, with many cases that are ultimately resolved by arbitration having started off in the courts, whether in the Sheriff Court or the Court of Session.

The 1996 Act introduced a statutory right to adjudication. Adjudication was not a creation of the 1996 Act; it had been available in Scotland in relation to certain disputes arising under the DOM/1/Scot and DOM/2/Scot forms of sub-contract and the With Contractor's Design versions of the Scottish Building Contract. Finally in this chapter we will consider Alternative Dispute Resolution.

## 14.2 *Arbitration*

### 14.2.1 What is arbitration?

Historically, most building contracts have contained an arbitration clause whereby parties to the contract may seek to have their dispute referred to the private determination of an appointed individual, an arbiter.

This is instead of seeking resolution through the courts. Arbitration is a formal alternative to litigation. It requires the agreement of both parties to the relevant contract. This agreement, if insisted upon, excludes from the jurisdiction of the courts any right to deal with the merits of the dispute, see *Hamlyn & Co* v. *Talisker Distillery* (1894).

Under the Scots law of arbitration, all powers and duties of the arbiter are derived from the contract itself and that which is express and implicit therein. There has been little statutory intervention. Arbitration law in Scotland is therefore based principally on common law, unlike the position in England.

**Perceived advantages**

Arbitration, as a method of resolving construction disputes, has often been preferred to litigation because of certain perceived advantages.

- It is private.
- It involves determination by an expert in the field.
- It has been considered, historically, to be less expensive and speedier.

Unlike litigation, where facts and arguments are aired in open court, arbitration is conducted in private. The advantage of having a dispute resolved by an expert in the field of construction, rather than a lawyer, continues to be the principal attraction of arbitration. The considerations of expense and speed have not been perceived to be advantages of the arbitration process in recent years. Sir Michael Latham, in his 1994 report *Constructing the Team,* said

> 'There is considerable dissatisfaction with arbitration within the construction industry because of its perceived complexity, slowness and expense.'

**The laws which apply**

This section is concerned with the Scots law of arbitration and relates to arbitration conducted in Scotland applying Scottish procedure. The questions at issue in such arbitrations may be determined by a foreign system of law. It is important not to confuse the law governing the arbitration procedure and the 'proper law' which is applied to determine the substantive matters at issue between the parties. These two need not be the same. However, in both cases, the parties themselves may agree which law to apply. If there is no agreement, an arbitration held in Scotland will be governed by Scots procedure. See *James Miller and Partners Ltd* v. *Whitworth Street Estates (Manchester) Ltd* (1970). The 'proper law' of the arbitration will be determined by normal private international law rules.

### 14.2.2 How do you choose arbitration?

Arbitration is a contractual process. The law presumes that the courts will settle any claims unless parties have agreed otherwise by what we shall call 'the arbitration agreement'. This can be done in building contracts in one of two ways. Most commonly, the building contract itself may contain an arbitration agreement. Alternatively, parties may at the time a dispute arises enter into a separate arbitration agreement providing that an arbiter shall be appointed to determine the dispute. This agreement can be specific to the dispute in question or general in that it covers all disputes which may be referred to the arbiter under a building contract.

The arbitration agreement requires parties to

(1) submit certain matters to an arbiter or arbiters;
(2) bind themselves to abide by his or their decision; and
(3) concur in the appointment of an arbiter or arbiters when called upon to do so, or the appointment of such arbiter or arbiters by another, identified in the contract.

### 14.2.3 Appointment of the arbiter

The JCT 98 arbitration agreement at clause 41B, as amended by the Scottish Building Contracts, requires the reference of a dispute to an arbiter, as agreed by the parties within 14 days after notice to concur in the appointment of such a person has been given, or as may be appointed by one of the three office bearers named in Appendix II. If the appointing office bearer is not identified the Chairman or Vice Chairman of the Scottish Building Contract Committee makes the appointment.

The Arbitration (Scotland) Act 1894 abolished the common law rule that an arbitration agreement was invalid if it did not name the arbiter. It further provided that an arbitration agreement would not be invalid simply by virtue of a third party appointer of the arbiter being named in the arbitration agreement.

The 1894 Act also provided that on refusal by one party to agree to an appointment or to a mechanism for so doing, the court could appoint the arbiter. This may be done by application either to the Court of Session or to a Sheriff in a Sheriffdom to whose jurisdiction the respondents are subject.

### 14.2.4 Powers of the arbiter

The arbitration agreement provides the arbiter with his powers and duties. Some agreements are very wide as to the express powers given to the arbiter. Others are in the briefest of terms. In either of these circumstances parties must look also to the powers implied by the common law. It has been held that an arbiter has no implied power at common law to award damages (see *Aberdeen Railway Co* v. *Blaikie Bros* (1853); or to award interest from a date prior to the issue of his award (see *Farrans (Construction) Ltd* v. *Dunfermline DC* (1988) and *John G McGregor (Contractors) Ltd* v. *Grampian Regional Council* (1991). There is also some doubt as to whether he has an implied power to make interim awards, see *Lyle* v. *Falconer* (1842). For this reason many arbitration agreements provide the arbiter with such powers. This is often done by reference to a set of arbitration rules published by any one of a number of professional bodies, e.g. The Chartered Institute of Arbitrators (Scottish Branch) or The Law Society of Scotland.

If the arbitration agreement is lacking in such detail, it is open to parties to enter into a supplementary arbitration agreement to deal with such matters. This often occurs at the time of a dispute arising.

The arbitration agreement may provide that both parties consent to registration of any arbiter's decree for preservation and execution in the Books of Council and Session. This is considered below in section 14.2.6. This provision is more commonly found in deeds of submission.

### 14.2.5 What may be referred to arbitration?

As the purpose of arbitration is to adjudicate upon the questions in issue between the parties, there must be certain matters at issue between the parties before a reference can be made to arbitration.

One must first examine the terms of the arbitration agreement to determine whether any issue that has arisen is covered by the agreement and therefore is capable of being referred to arbitration. Historically, the scope of the arbitration agreement was construed strictly, on the basis that for parties to oust the jurisdiction of the court this had to be well spelt out. Arbitration agreements in most modern standard forms of building contracts are wide enough to cover all disputes which arise out of the contract.

The JCT 98 arbitration agreement at clause 41B, as amended by the Scottish Building Contract, refers to arbitration

> 'any dispute or difference between the employer and the contractor arising during the progress of the works or after completion or abandonment thereof in regard to any matter or thing whatsoever arising out of the contract or in connection therewith ...'.

It then specifically describes certain types of dispute which fall within the definition. The words 'arising out of ... or in connection ... with' cover not only claims under the contract but also claims for damages for breach of contract. Disputes relating to liquidated and ascertained damages arising from a contract will normally be covered by an arbitration agreement without express provision.

The scope of disputes covered by the arbitration agreement does not imply that the arbiter has sufficient powers to deal with such disputes. An arbiter, for example, may have had disputes on breach of contract validly referred to him but have been given no power to award damages. In the absence of an express power to award damages, an arbiter has no implied power to do so. In the absence of any such express power the arbiter may decide on whether or not there is a liability for damages but it must be left to the court to assess and award such damages. See *North British Railway Co* v. *Newburgh & North Fife Railway Co* (1911).

The wording of the arbitration agreement is, therefore, critical in determining what may be referred to arbitration, and what the arbiter has power to deal with. As will be explained later, if the arbiter exceeds the limits of his appointment his whole award is subject to challenge in the courts.

To refer a matter to arbitration, a dispute, as described in the arbitration agreement, must have arisen. The arbitration agreement may prevent arbitration upon certain matters until after practical completion so it must be scrutinised closely.

There may be a situation where one of the parties to the arbitration agreement does not accept that a dispute, as defined in the arbitration agreement, has arisen. The courts may then need to decide whether a dispute does exist, see *Mackay & Son* v. *Leven Police Commissioner* (1893).

There must have been a communication by the defender (or respondent) which rejects or disputes the claim. A matter is no longer in dispute if it has been determined by a court or other legal tribunal.

In the face of a refusal to submit to arbitration, the courts may become involved. This may be where the pursuer (or claimant) seeks to have the court order the defender to submit a particular dispute to arbitration. Alternatively, the pursuer may not wish to invoke the arbitration clause as he believes there is no dispute and the court should grant summary decree. In *Redpath Dorman Long Ltd* v. *Tarmac Construction Ltd* (1982) the defenders, as main contractors, sought to have a claim by a sub-contractor referred to arbitration. The pursuers maintained that there was no dispute and sought summary decree. The court found that failure to issue a notice of set-off as required by the sub-contract meant there was no defence (and so no dispute) in respect of part of the claim, which could not, therefore, be referred to arbitration. Summary decree was pronounced.

However, the courts have taken a very broad view of what constitutes a dispute. Current authorities suggest that the court shall, if there is any suggestion of a dispute, refer the matter to arbitration to enable the arbiter so to determine, see *Halki Shipping Corporation* v. *Sopex Oils Ltd* (1997). A dispute on an adjudicator's decision is an exception to this principle, see *Macob Civil Engineering Ltd* v. *Morrison Construction Ltd* (1999).

### 14.2.6 How do you commence an arbitration?

The arbitration agreement may provide how arbitration proceedings are to be commenced. In general the reference of a dispute to arbitration may be done by serving a notice to refer to arbitration or a reference from court proceedings to arbitration.

#### Notice to refer

In the case of a notice to refer, the arbitration agreement may prescribe its terms or details. The notice must describe the dispute which it is sought to refer and the mechanism whereby an arbiter is to be selected. Commonly, the party issuing the notice provides two or three names of proposed arbiters and calls upon the other party to agree one of these names or propose another individual within, say, 14 days, failing which the party seeking arbitration will invoke the appointment mechanism within the arbitration agreement.

The notice to refer a dispute to arbitration, or 'preliminary notice', is of importance because it defines the dispute which is to be remitted to the arbiter and his jurisdiction. It is, further, the equivalent of court proceedings in interrupting the running of the prescriptive periods and determining whether a claim is time barred. Prescription is considered above in Section 8.9.

It is essential to include in the preliminary notice all matters which the claimant requires the arbiter to determine. It is possible to serve further notices but the date of each notice will be the date of interruption of the prescriptive period in respect of the matters contained therein.

The nature of the dispute must be adequately set out in the preliminary notice. It is not enough to refer in the preliminary notice to a separate claim document, see *Douglas Milne Ltd* v. *Borders Regional Council* (1990). If the preliminary notice does not specify the claim, the date of interruption of the prescriptive period will be the date when the claim is made in the arbitration, see *John O'Connor Plant Hire* v. *Kier Construction Ltd* (1990).

## Deeds of submission

A notice to refer to arbitration may, sometimes, be followed by a deed of submission signed by both parties which outlines the dispute being remitted and identifies the agreed arbiter. This formalises the submission but is not a necessary requirement. The opportunity, however, may be taken by virtue of such a deed of submission to extend the arbiter's powers in addition to those contained in the original arbitration agreement and to stipulate whether a particular set of arbitration rules are to apply.

The deed may also contain a provision that allows enforcement of any decree arbitral by containing agreement to register any decree of the arbiter in the Books of Council and Session together with the deed of submission. This avoids the need to revert to court to seek an award implementing a decree arbitral. The Books of Council and Session are a division of the Registers of Scotland. The equivalent of a court decree is obtained by registering the decree arbitral and obtaining an extract for enforcement purposes, provided there is a decree upon which diligence can be carried out.

The document which contains the consent to register must have formal validity, as must the decree arbitral.

## Court proceedings

A party may, instead of serving a preliminary notice, raise a court action and then, provided the plea is taken timeously, have the action sisted (put on hold) for arbitration. In these circumstances, the writ defines the scope of the arbiter's remit unless a separate preliminary notice or deed of submission is entered into which covers this.

The intention to invoke an arbitration clause and remit a dispute to arbitration does not preclude the issuing of court proceedings. However, if court proceedings are initiated, either party to the proceedings must invoke the arbitration clause if they intend to do so. A plea to sist an action pending arbitration may be competently made at any time before the closing of the record. No inference of abandonment of that right is derived from parties having engaged in the ordinary procedures of litigation before the plea is taken. For a party to be held to have abandoned the right to go to arbitration, their actings and/or their failure to act, must be inconsistent with the intention to insist upon arbitration, see *Presslie* v. *Cochrane McGregor Group Ltd* (1996). The arbitration provisions are invoked by including a plea in the pleadings that the action should be sisted pending the outcome of the arbitration. The courts have no discretion if the plea is insisted upon and the right to arbitration has not been waived.

A court action may also be raised in parallel with serving a preliminary notice although it will need to be sisted shortly thereafter. This may be done for a number of reasons. If the arbiter has no power to award damages or interest prior to the date of his award, it may be necessary to have recourse to the court. To protect the position it may, therefore, be sensible to raise court proceedings at the outset and sist them. These issues may then be dealt with after the arbitration is completed.

It may also be thought desirable to seek such protective measures as the court may allow, pending the outcome of the arbitration, such as arrestment, inhibition or interdict. The existence of arbitration proceedings does not prevent an action replicating the arbitration claim being raised in court to obtain such interim measures. The court action is therefore preserved but the dispute continues to be resolved by arbitration, see *Graeme Borthwick Ltd* v. *Walco Developments (Edinburgh) Ltd* (1980).

### 14.2.7 Formalities of the arbiter's appointment

Once the arbiter has been agreed or nominated, he must ensure he has a contract with the parties to the process. This is normally done by the arbiter, on being approached, accepting office subject to his terms being accepted by both parties. These will cover his fees and any other ancillary matters he may wish to raise, such as the appointment of a clerk.

#### The arbiter's clerk

There has been a tradition in Scottish building contract arbitrations that a clerk, who is a lawyer, is appointed by a 'non legal' arbiter to assist him in the framing of interlocutors (orders) and decisions and to advise him on the law where necessary. This may involve the clerk attending all hearings, having custody of the process and issuing all documents, including the orders of the arbiter. He will also hold any tender (private offer to settle) submitted by one of the parties. In recent times this tradition has been departed from on occasion, primarily, it is to be presumed, due to cost considerations. An option is for a clerk to be identified and agreed, but to be called upon only when and if the arbiter requires.

### Ensuring impartiality

As arbitration is within a contractual framework, parties are free to choose or agree upon any individual to act as arbiter. However, as it is a judicial process, it is implicit in any appointment that the arbiter must be unbiased and have no financial or personal interest in the dispute or outcome.

Any person selected to act as arbiter must disclose to the parties to the appointment any interest or connection he has with the parties or the dispute. This enables parties to determine whether to waive any objection they might otherwise have as to the validity of his appointment, see *Edinburgh Magistrates* v. *Lownie* (1903). If one of the parties considers the arbiter to be disqualified on the grounds of his partiality, interdict may be sought to prevent him taking up office.

### Termination of the appointment

The appointment of an arbiter may be terminated by the agreement of both parties. Where an arbiter has been appointed, but the claimant does not progress the arbitration, the issue may be whether parties have, by their inaction, abandoned the contract. The difficulty arises as to what constitutes implied consent to abandonment by the respondent and to inaction implying abandonment by the claimant, see *Paal Wilson & Co A/S* v. *Partenreederei Hannah Buumenthal* (1983).

Failure of the arbiter to progress the arbitration may, in extreme circumstances, justify an order being sought in court to order him to perform the duties of his office. His failure to obey this order might then lead to an argument that the contract was frustrated. A submission to arbitration may be frustrated if the arbiter dies or becomes mentally or physically incapable.

### Liability of the arbiter

There is no Scottish decision which deals with the liability of the arbiter for damages. As an arbiter exercises a judicial function, it is considered that any error of judgment would not make him liable in damages to the parties. This rule is also likely to be applied in relation to any potential liability for negligence. However, see *Arenson* v. *Arenson* (1977), where there has been gross misconduct on the arbiter's part, for example bribery or corruption, he will have a potential liability.

**The arbiter's fees**

The arbiter is entitled to be remunerated for his services in the same way as those engaged in other employment. The arbiter normally ensures that the level of his fees is agreed at the outset. If not, however, he would be entitled to be paid *quantum meruit* (a reasonable amount for the work done). Liability for the arbiter's fees and expenses, in the absence of agreement or order to the contrary, is shared by the parties jointly and severally.

### 14.2.8 The procedure of the arbitration

Subject to contractual restraints, the arbiter is master of his procedure and can determine whatever procedure he considers appropriate for the effective conduct of the case. He is acting in a judicial capacity and must therefore adequately inform his mind as to the nature and merits of the dispute. This is subject to him ensuring that the rules of natural justice are followed, allowing each party an equal opportunity to prepare and present their case. He must be impartial. He cannot resign without cause.

The arbitration agreement, or the deed of submission, may incorporate by reference standard arbitration rules of procedure. These are produced by a number of bodies. It may alternatively have set out its own detailed rules of procedure. In the absence of either, the arbiter must apply his own rules in the contractual framework within which he operates.

The arbiter may determine that the dispute is capable of resolution on 'documents only', without oral hearing or submission. This will ordinarily require the consent of both parties. Alternatively, he may follow the traditional court approach to pleadings and order a statement of claim and answers and thereafter a period of adjustment by both parties of these pleadings. Further procedure by way of preliminary legal hearing, debate or a proof, where witnesses give evidence, will be determined by him, normally following representations of the parties.

On the route to final determination of the dispute, certain issues may emerge which require to be dealt with by the arbiter. These include applications for recovery of documents, enforcement of interim or partial awards and orders for the citation of witnesses. As the arbiter is a private judge, he has no power to enforce any orders he may make. He is reliant on the court to lend authority for this.

In the case of recovery of documents, an application will normally be made to the arbiter listing the documents or other property which it is sought to recover and citing the legal justification therefor. To the extent that the arbiter approves this, he issues an order to this effect and application can then be made to the court for an order enforcing the same, see *Crudens Ltd Petitioners* (1971). An order for attendance of witnesses at a proof or hearing may be made by the court on application by a party, following an order of the arbiter recommending that such an application is granted.

### 14.2.9 The award

It is common practice for the arbiter to issue proposed findings of fact and his award in draft. Parties are then asked to comment on any errors or inconsistencies therein.

The award takes effect when delivered to the clerk or the parties. An arbiter may withhold delivery of his award until his fees and those of his clerk have been paid. The award is known as the decree arbitral. The award must be clear and certain. It must be capable of enforcement, whether in respect of sums of money or other matters. It is advisable that the award be signed by the arbiter in the presence of a witness, see *McLaren* v. *Aikman* (1939) and the Requirements of Writing (Scotland) Act 1995. This enables registration in the Books of Council and Session if parties have so consented.

Once the arbiter has issued his final award he is *functus officio* – he has no further contractual authority, he cannot alter or correct the award, and he has no jurisdiction to enforce the award. This may be done by a court action for implement of the award, or for payment where it is a monetary award. The court will, unless there is objection to the validity of the award, pronounce a decree for implement. Alternatively, enforcement can be effected by parties signifying their consent to the registration of the decree arbitral in the Books of Council and Session for preservation and execution. This consent can be given in the signed arbitration agreement or a deed of submission and the agreement to that effect can be endorsed by the parties on the decree arbitral itself.

### 14.2.10 Recourse to the court

There has already been described above the types of arbitral orders on which the court's authority is sought during the conduct of the arbitration.

The court has no general power at common law to review the course or conduct of the arbitration. Only where the arbiter plans to or has been called upon to exceed his powers or jurisdiction, may the court interdict him from so doing.

### 14.2.11 Stated case

Section 3 of the Administration of Justice (Scotland) Act 1972 introduced, for the first time in Scotland, a provision for the arbiter to apply, by way of a stated case, to the Court of Session for an opinion on a question of law. He may do so on application by one of the parties to the arbitration and shall do so on order of the court. An application for the arbiter to state such a case may be made at any stage in the arbitration. This means that it must be made before the final award, see *Fairlie Yacht Slip Ltd* v. *Lumsden* (1977). This has led to the common practice of an arbiter being requested by the parties to first issue his proposed final award in draft form. The procedure for stating a case is regulated by the Court of Session Rules. If the arbiter refuses to state a case, a party may apply to the court for an order ordaining him to do so. The strict time limits set down in these rules must be adhered to. An appeal to the House of Lords from the Court of Session under s. 3 of the 1972 Act is incompetent, see *John G McGregor (Contractors) Ltd* v. *Grampian Regional Council* (1991). The power to require the arbiter to state a case may be expressly excluded by parties in the arbitration agreement.

### 14.2.12 Reduction of award

An arbiter's award may be challenged in the court under the Articles of Regulation 1695 or as being contrary to common law. Under the 25th Act of these Articles, an action of reduction of an arbiter's award is possible on the grounds of corruption, bribery or falsehood of the arbiter. More commonly, a challenge may be mounted at common law on certain very limited grounds. This is done by raising an action of reduction in the Court of Session.

An order for reduction sets aside the award which has been challenged. The court cannot, however, decide on the dispute and the outcome is, in theory, another arbitration with a new arbiter.

Apart from the invalidity of the award due to procedural failings, namely, failure to sign the award properly or failure to deliver the award, there are the following possible grounds of reduction.

#### The award exceeds the arbiter's jurisdiction

This may be because the arbiter deals in the award with matters not expressly submitted to him by the reference. Alternatively the arbiter may have exceeded his powers, express or implied.

#### The award does not exhaust the submission

Failure to pronounce on all issues submitted to the arbiter is fatal to his award. The award cannot stand if challenged on any of the issues unless there is provision in the arbitration agreement to issue part awards. This does not extend to a failure on the arbiter's part to deal with the question of expenses, which are considered incidental to the process.

#### Where the award is uncertain

If the award is so uncertain in its terms as to be unenforceable, it can be reduced.

### Breach of natural justice

Where the arbitration has been conducted contrary to the rules of natural justice the award can be reduced. This has sometimes been classified as misconduct of the arbiter. This situation arises where, for example, the arbiter has not given the parties equal and fair hearings, or where he refused to allow one party to respond to the other party's case. These are classified as breaches of the implied duties of the submission.

## 14.2.13 Judicial review

It is now accepted that judicial review is the more appropriate means to challenge the actings of an arbiter. Thus, in circumstances where the common law action of reduction was formerly used (as in the instances described above), it is now appropriate to proceed by way of judicial review, see *Kyle & Carrick DC* v. *A R Kerr & Sons* (1992) and *West* v. *Secretary of State for Scotland* (1992).

However, it is important to emphasise that the purpose of judicial review is to exercise supervisory jurisdiction. That is, to ensure that a person or body does not exceed or abuse their jurisdiction, power or authority or fail to do what the jurisdiction, power or authority requires. It is not competent for the court to review the acts or decisions on their merits, nor may it substitute its own opinion for that of the person or body, see *West*. The remedy of judicial review is not available as some kind of back door means of reviewing the merits of an arbiter's award. If it is considered that the decision contains an error in law then the remedy is by way of stated case under the 1972 Act.

If the arbiter has acted within the boundaries of his discretion as arbiter, and it cannot be inferred that no reasonable arbiter would have so acted, the court will not interfere with the decision of the arbiter, see *Shanks & McEwan (Contractors) Ltd* v. *Mifflin Construction Ltd* (1993).

## 14.2.14 UNCITRAL Model Law

The UNCITRAL Model Law of Commercial Arbitrations applies to international commercial arbitrations conducted in Scotland by virtue of Schedule 7 to the Law Reform (Miscellaneous Provisions) (Scotland) Act 1990.

The Model Law applies to all arbitrations conducted in Scotland where the parties have their places of business in different states or if the work is to be performed in a different state. An arbitration is conducted in Scotland for these purposes where the seat of the arbitration is said to be in Scotland. The definition of state has Scotland as a state in its own right. This results in the application of the Model Law to cross Scottish/English border arbitrations unless parties agree otherwise. Parties may contract in or out of the Model Law.

The Model Law has a dual role. First, it alters the substantive law. Thus, for example, the possibility for parties to have recourse to the courts is greatly reduced. There is no right to state a case under the Model Law. Secondly, the Model Law contains procedural rules relating to the running of the arbitration.

### 14.2.15 The future of arbitration in Scotland

In March 1996 the Scottish Advisory Committee on Arbitration Law chaired by Lord Dervaird ('the Dervaird Committee') issued its report on the question of legislation for domestic arbitrations in Scotland. This report made a number of recommendations and also incorporated a draft Arbitration (Scotland) Bill. The recommendations included the making of statutory provisions:

- conferring power on an arbiter to award interest
- conferring power on an arbiter to award damages
- conferring power on an arbiter to continue with an arbitration notwithstanding failure by a party to appear at a hearing or comply with an order by the arbiter
- repealing the rights of appeal by way of a stated case under s.3 of the Administration of Justice (Scotland) Act 1972.

While some reform of the Scottish Law of Arbitration is likely in the near future, no timetable for this has been set and it is not clear whether any such reform will be as recommended by the Dervaird Committee or in the form of a more radical codification.

## *14.3 Litigation*

### 14.3.1 Sheriff Court and Court of Session

In the absence of an arbitration agreement, or where parties have waived the arbitration agreement, the courts in Scotland have, traditionally, resolved building contract disputes. This can be in the Sheriff Court, which is the local court for each area, or the Court of Session, which is based in Edinburgh. Typically, the smaller to medium sized claims are pursued in the Sheriff Court. The selection of which Sheriff Court the action can be pursued in depends upon the rules of jurisdiction. For example, the Sheriff Court of the defender's place of business, or that of the pursuer's business address where the sums sued for are to be paid, are two possible grounds of jurisdiction. The Civil Jurisdiction and Judgements Act 1982 sets out the various grounds of jurisdiction. Parties may agree in the building contract which courts have jurisdiction.

Claims with a monetary value of up to £1500 must be pursued in the Sheriff Court. Apart from this, the Court of Session and the Sheriff Court have concurrent jurisdiction in respect of monetary claims.

Within the Court of Session, a choice needs to be made by the pursuer as to whether to initiate proceedings in the Commercial Court or under the ordinary procedure. The rules provide for a more flexible procedure to be adopted by the Commercial Court. It is considered that most building contract claims fall within the definition of commercial actions, which entitles the use of the Commercial Court.

The rules of procedure of the Sheriff Court and the Court of Session are, respectively, set out in the Ordinary Cause Rules 1993 and Schedule 2 of the Court of Session Act 1994, both as amended from time to time.

Clause 41C of JCT 98, as amended by the Scottish Building Contract, introduced, for the first time, an option to select litigation rather than arbitration as the method of resolving disputes under the contract.

### 14.3.2 The procedure

A detailed examination of court procedure is beyond the scope of this book, however, the important features of Court of Session actions and of Ordinary Cause actions in the Sheriff Court are set out below.

#### The initiating writ or summons

In the Court of Session, an action for payment is initiated by a summons whether in the Commercial Court or not. An ordinary cause in the Sheriff Court is initiated by initial writ. The effect of service of the writ or summons is to interrupt the running of the prescriptive period within which claims may be pursued.

#### Sheriff Court

In the Sheriff Court a defender has 21 days from service of the initial writ to defend the action. If the defender does not lodge a notice of intention to defend within that period, the pursuer may seek a decree.

If a notice of intention to defend is lodged, the defender must lodge defences 14 days after the 21 day period of notice has expired. There then follows a fixed period where each party expands on their pleadings to respond to the other side's case and focus the issues between them. This is known as the period of adjustment. If the defender also has a claim or right of set-off, a counterclaim setting this out may be lodged and dealt with at the same time. After the period of adjustment, the court will hear both parties at an options hearing to determine further procedure. The pleadings consisting of the initial writ, defences and counterclaim, where appropriate, and the adjustments thereto are put together in a document known as the 'closed record' when the permitted adjustment period expires.

Further procedure may take the form of a debate, which is a hearing to deal with legal issues. This may resolve the whole action or it may lead to refinement of a party's case. A proof or a proof before answer may be ordered after, or instead of, a debate. A proof is to try the factual issues of the case by hearing evidence from witnesses. Where legal arguments have been reserved and are to be heard at the conclusion of the evidence, the hearing is known as a proof before answer.

Incidental applications for matters such as the recovery of documents or interim decree are done by way of written motion.

### The Court of Session

The ordinary procedure of the Court of Session is similar to that of the Sheriff Court. Rather than lodge a notice of intention to defend, the party proposing to defend the action must enter appearance once the Summons is called. The period for lodging defences is seven days from the date of calling. A debate is described as a 'procedure roll' hearing.

The Commercial Court of the Court of Session has a somewhat more flexible procedure. There is no automatic right to adjustment of pleadings. The commercial judge is pro-active in his handling of the case. A case will tend to be dealt with by the same judge throughout. The case first calls in front of this judge for a preliminary hearing within seven days of the expiry of the period for lodging of defences. Thereafter, there will follow a series of hearings designed around the requirements of the case. A proof, proof before answer or debate, or some alternative procedure may follow. Incidental applications are made by written motion.

## 14.3.3 Protective measures

The right to seek protective measures to secure the claim or the subject matter of the dispute, pending the outcome of proceedings, may be exercised at the outset of Sheriff Court or Court of Session proceedings, or at a later date by motion.

The principal protective measures are the rights to arrest or inhibit pending the outcome of the action (known as warrant to arrest and inhibit on the dependence). These are currently granted automatically, when lodging a Court of Session summons or counterclaim, if applied for on the face of the summons or counterclaim. In the Sheriff Court a warrant to arrest is granted automatically, if applied for when lodging the initial writ.

An inhibition can only be granted by the Court of Session. This means that where inhibition is sought in a Sheriff Court action, a separate application by means of petition to the Court of Session is required.

## 14.3.4 Appeals

In the Sheriff Court parties can appeal to the Sheriff Principal or to the Court of Session. Leave (or permission) to appeal is necessary in certain circumstances. A further appeal from the Sheriff Principal to the Inner House of the Court of Session may also be competent. The appeals procedure in the Sheriff Court is regulated by the Sheriff Courts (Scotland) Act 1907 as amended by the Sheriff Courts (Scotland) Act 1971.

The appeal function of the Court of Session is exercised by the Inner House. If the appeal is from a Court of Session judge's decision, it is known as a reclaiming motion. As with the Sheriff Court, leave to appeal may be required in certain circumstances.

Finally, an appeal from the Inner House of the Court of Session to the House of Lords may be competent. The appeals procedure in the Court of Session, including the method by which an appeal to the House of Lords can be taken, is regulated by the Rules of the Court of Session and the Court of Session Act 1988.

## 14.4 *Adjudication*

The 1996 Act introduced a statutory right to adjudication into all construction contracts entered into on 1 May 1998 and thereafter. Adjudication for these purposes may be defined as a quick enforceable interim decision which lasts unless or until the dispute is otherwise determined by arbitration or litigation. It does not involve final disposal of the dispute unless parties agree to accept it as such.

Sections 104 to 108 of the 1996 Act require parties to include certain terms in their contract regarding dispute resolution. This requires every construction contract to contain four essentials. These are:

(1) provision whereby either party may refer any dispute or difference which arises under the contract to adjudication;
(2) a stipulated procedure for referral whereby a decision is given within a 28 day period from the date of referral, which can be extended to six weeks by the referring party or by any period with the agreement of both parties;
(3) provision that the decision is binding albeit on an interim basis; and
(4) provision that the adjudicator is not liable unless in bad faith.

If any one of these essentials is missing from the contract, a statutory scheme, contained within The Scheme for Construction Contracts (Scotland) Regulations 1998 ('the Scheme') is applied to the contract in question and has the effect of incorporating the provisions of the Scheme as terms of the contract, see s. 114(4) of the 1996 Act. The Scheme is therefore invoked where the contract is silent on adjudication or does not contain all the essentials referred to above.

The standard forms of building and engineering contracts have amended their terms to include an adjudication clause. JCT 98, as amended by the Scottish Building Contract, has introduced its own detailed adjudication provisions, see clause 41A.

### 14.4.1 Who may be an adjudicator?

Any person agreed upon by parties may be an adjudicator. In the absence of such agreement, an adjudicator may be appointed by either an adjudicator nominating body named in the contract or, if silent, one of the adjudicator nominating bodies which holds itself out to be such. There are no restrictions on who may be an adjudicator. Professional bodies such as ICE, RIAS, RICS and CIOB are just some of the bodies who have trained and certified adjudicators.

### 14.4.2 What may an adjudicator do?

He shall decide the disputes referred to him. In doing so, he may take the initiative in ascertaining the facts and the law. His powers are wide. They may be circumscribed by the adjudication clause in the contract. His decision is binding unless or until parties take the matter to arbitration or litigation and the dispute is otherwise dealt with there. Parties may agree that the adjudication award is to be final and binding of parties' rights and obligations. The adjudicator is not restricted, unless the contractual adjudication clause says so, as to the nature of remedy he may grant.

### 14.4.3 How do you go to adjudication?

This is done by issuing a notice to refer a dispute to adjudication. The form of this notice and procedure thereafter is regulated by the contractual adjudication clause or the Scheme, whichever may apply.

### 14.4.4 Enforcement of adjudicator's decision

The 1996 Act does not deal with enforcement. It simply provides, at s.108(6), that the Scheme may provide the court with powers on adjudication and make provisions relating to the enforcement of the adjudicator's decision. The Scheme does not, in its current form, provide the court with such powers. It does, however, provide (at Part II of the Schedule to the Scheme, paragraph 24) that if one of the parties or the adjudicator wishes to register the decision for execution in the Books of Council and Session any other party shall, on being requested to do so, consent forthwith to such registration. To date, enforcement of this has been by petitioning the Court of Session (under judicial review procedure) for an order requiring specific performance.

The adjudication clause in the contract may provide for enforcement. This may also include a requirement to give consent to any adjudicator's decision being registered in the Books of Council and Session for execution. JCT 98, as amended by the Scottish Building Contract, provides for this consent to registration in the separate SBCC Adjudication Agreement.

Alternatively, the adjudication clause may provide that the adjudicator's decision is to be complied with, i.e. it is a contractual obligation. Finally a form of 'decree conform' by the court, enabling enforcement of the adjudicator's award, may be sought.

### 14.4.5 Rights of review

There are no provisions for any review of the adjudicator's award until it is referred to arbitration or litigation, whichever is the ultimate means of dispute disposal. The Scottish courts have shown a reluctance to interfere in a party's contractual arrangements as to how to resolve disputes. Any right of review is therefore likely to be restricted to procedural irregularity which has resulted in a defective decision. This will include where the decision has been arrived at contrary to the rules of natural justice. In addition, where the adjudicator has acted outwith the scope of his authority, such a decision may be open to challenge.

## *14.5 Alternative Dispute Resolution*

Only a very small percentage of building contract cases go the full way in arbitration or litigation. The vast majority settle in the run up to the full hearing or proof. The settlement is often the result of negotiations at client or lawyer level. Over the past few years a more formal process of negotiation of a case to settlement has developed. This is known as Alternative Dispute Resolution or ADR.

ADR can take many forms. It may be initiated as an alternative to litigation or arbitration or it may be conducted in parallel with the process of litigation or arbitration.

Its aim is obvious – to achieve a negotiated settlement in an economic and effective manner. The means are various, including mediation and mini-trial. As the process is consensual, a reference to ADR is non-binding. Parties may withdraw at any time until they have formalised the terms of any settlement reached.

A common feature of ADR, although not essential, is the appointment of a third party 'neutral' to conduct or facilitate the process. This individual may be appointed as a mediator, where his role is to mediate between the two or more parties to facilitate settlement. This often involves a form of shuttle diplomacy with the mediator representing the parties' positions to each other, only bringing them together once common ground has been found.

At a mini-trial, parties put forward, in a considerably condensed form, their cases. It is normally chaired by a neutral with one senior executive from each side. Following the hearing this body is required to agree on the outcome, analysing the strengths and weaknesses of each party's case.

A number of organisations have trained neutrals, or mediators, available for appointment.

One of the advantages of ADR, apart from speed and being relatively inexpensive, is that the settlement can cover any number of matters. It is not constrained by the issues in dispute.

The timing of successful ADR is all important. If a reference is sought too early, the issues may not be sufficiently well prepared or focused to form a background for discussion; too late and the parties have become entrenched in their attitudes. A will to resolve the matter through the process is necessary before progress can be made.

ADR can be initiated at any time by either of the parties inviting the other to accede to the process, or by having one of the ADR bodies opening up the dialogue. Parties may be represented by lawyers through the process, but it is by no means essential. Normally the costs of the whole process are shared equally by the parties.

Appendix
# Useful Addresses

**Association of Consulting Engineers (ACE)**
Alliance House
12 Caxton Street
London SW1H 0QL
tel: 0171 222 6557
fax: 0171 222 0750
e-mail: consult@acenet.co.uk
website: www.acenet.co.uk

**Civil Engineering Contractors Association (CECA)**
Construction House
56–64 Leonard Street
London EC2A 4JX
tel: 0171 608 5060
fax: 0171 608 5061

**Institution of Civil Engineers (ICE)**
1 Great George Street
London SW1P 3AA
tel: 0171 222 7722
fax: 0171 222 7500
e-mail: ice.org.uk
website: www.ice.org.uk

**Royal Incorporation of Architects in Scotland (RIAS)**
15 Rutland Square
Edinburgh EH1 2BE
tel: 0131 229 7545/7205
fax: 0131 228 2188
E-mail: info@rias.org.uk
website: www.rias.org.uk

**RIAS bookshops:**
Edinburgh:
15 Rutland Square
Edinburgh EH1 2BE
tel: 0131 229 7545
fax: 0131 228 2188
website: www.rias.org.uk

Glasgow:
Mackintosh School of Architecture
The Bourdon Building
177 Renfrew Street
Glasgow G3 6RQ
tel: 0141 332 9414
fax: 0141 332 9252

**Royal Institution of Chartered Surveyors (RICS)**
9 Manor Place
Edinburgh EH3 7DN
tel: 0131 225 7078
fax: 0131 226 3577
website: www.rics.org.uk

**Scottish Building Contracts Committee (SBCC)**
27 Melville Street
Edinburgh EH3 7JF

# Table of Cases

The following abbreviations of Reports are used:

| | |
|---|---|
| **AC** | Law Reports, Appeal Cases |
| **All ER** | All England Law Reports |
| **App Case** | Law Reports, Appeal Cases |
| **BLR** | Building Law Reports |
| **CLR** | Commonwealth Law Reports |
| **CILL** | Construction Industry Law Letter |
| **Const LJ** | Construction Law Journal |
| **Con LR** | Construction Law Reports |
| **D** | Dunlop's Session Cases 1838–62 |
| **DLR** | Dominion Law Reports |
| **Ex** | Exchequer Reports |
| **F** | Fraser's Session Cases 1898–1906 |
| **GWD** | Greens Weekly Digest |
| **Hudsons BC** | Hudsons Building Contracts |
| **H&N** | Hurlstone & Norman |
| **IRLR** | Industrial Relations Law Reports |
| **KB** | Law Reports, Kings Bench |
| **M** | Macpherson's Session Cases 1862–73 |
| **Macq** | Macqueens Scotch Appeal Cases |
| **NSWLR** | New South Wales Law Reports |
| **PD** | Law Reports, Probate Division |
| **QB** | Law Reports, Queens Bench |
| **R** | Rettie's Session Cases |
| **Sh Ct Rep** | Sheriff Court Reports |
| **SLT** | Scots Law Times |
| **SCLR** | Scottish Civil Law Reports |
| **SLR** | Scottish Law Reporter |
| **SC** | Session Cases 1907– |
| **S** | P Shaw's Session Cases 1821–38 |
| **WLR** | Weekly Law Reports |

# Table of Statutes

# Table of Statutory Instruments

# Table of References to the JCT Standard Form of Building Contract (1998 Edition)

# Index